SOCIAL PERSPECTIVES IN THE 21ST CENTURY

AF758074

FROM HISTORICAL SOCIAL THEORY TO FOUCAULT

SOCIAL PERSPECTIVES IN THE 21ST CENTURY

JASON L. POWELL - SERIES EDITOR -

DEAN OF FACULTY, UNIVERSITY OF LIVERPOOL

Foucault: Issues and Legacy
Jason L. Powell (Author)
2012- November. ISBN: 978-1-62257-539-8 (Softcover)

Feminism
Jason L. Powell (Author)
2012- November. ISBN: 978-1-62257-540-4 (Softcover)

Habermas
Jason L. Powell (Author)
2012- November. ISBN: 978-1-62257-542-8 (Softcover)

Baudrillard and Postmodernism
Jason L. Powell (Author)
2013- January. ISBN: 978-1-62257-541-1 (Softcover)

Understanding Foucault: For Beginners
Jason L. Powell (Author)
2013. ISBN: 978-1-62417-195-6 (Softcover)

Understanding Power and Emotion: An Introduction
Jason L. Powell (Author)
2013. ISBN: 978-1-62417-200-7 (Softcover)

Understanding Risk and Trust: A Short Conceptual Examination
Jason L. Powell (Author)
2013- April. ISBN: 978-1-62417-202-1 (Softcover)

From Historical Social Theory to Foucault
Jason L. Powell (Author)
2013 - 3rd Quarter. ISBN: 978-1-62618-345-2 (Softcover)

SOCIAL PERSPECTIVES IN THE 21ST CENTURY

FROM HISTORICAL SOCIAL THEORY TO FOUCAULT

JASON L. POWELL

NOVINKA

New York

Copyright © 2013 by Nova Science Publishers, Inc.

All rights reserved. No part of this book may be reproduced, stored in a retrieval system or transmitted in any form or by any means: electronic, electrostatic, magnetic, tape, mechanical photocopying, recording or otherwise without the written permission of the Publisher.

For permission to use material from this book please contact us:
Telephone 631-231-7269; Fax 631-231-8175
Web Site: http://www.novapublishers.com

NOTICE TO THE READER

The Publisher has taken reasonable care in the preparation of this book, but makes no expressed or implied warranty of any kind and assumes no responsibility for any errors or omissions. No liability is assumed for incidental or consequential damages in connection with or arising out of information contained in this book. The Publisher shall not be liable for any special, consequential, or exemplary damages resulting, in whole or in part, from the readers' use of, or reliance upon, this material. Any parts of this book based on government reports are so indicated and copyright is claimed for those parts to the extent applicable to compilations of such works.

Independent verification should be sought for any data, advice or recommendations contained in this book. In addition, no responsibility is assumed by the publisher for any injury and/or damage to persons or property arising from any methods, products, instructions, ideas or otherwise contained in this publication.

This publication is designed to provide accurate and authoritative information with regard to the subject matter covered herein. It is sold with the clear understanding that the Publisher is not engaged in rendering legal or any other professional services. If legal or any other expert assistance is required, the services of a competent person should be sought. FROM A DECLARATION OF PARTICIPANTS JOINTLY ADOPTED BY A COMMITTEE OF THE AMERICAN BAR ASSOCIATION AND A COMMITTEE OF PUBLISHERS.

Additional color graphics may be available in the e-book version of this book.

Library of Congress Cataloging-in-Publication Data

From historical social theory to Foucault / editor, Jason L. Powell (University of Central Lancashire, UK).
pages cm
Includes bibliographical references.
ISBN 978-1-62618-345-2 (soft cover)
1. Social sciences--Philosophy. 2. Postmodernism. I. Powell, Jason L., 1971-
H61.F686 2013
300.1--dc23

2013007543

Published by Nova Science Publishers, Inc. † New York

CONTENTS

Chapter 1

INTRODUCTION

This book engages with what we understand by social theory and its rich and diverse perspectives. The informed use of social theory is part of the aim to critically question 'taken for granted assumptions', explicate theoretical interpretations of the social world and the place of human subjectivity within it. A 'theory' asks why a particular analogy is used to explain what is meant by society and relations within it and how the assumptions contained within social spaces influence our understanding of the position of individuals in historical and contemporary society. Social theories are important in establishing frameworks for understanding, interpreting human action and how social processes in everyday life are contested and negotiated, as well as the interplay between various 'levels' at which social relations take place: micro/macro and agency/structure. In approaching the writing of this book, there are two key questions that need to be flagged up: first, how does one make sense of the complicated nature of social theory? If we use a metaphor, a 'life-jacket' is needed which enables a committed reader to 'swim' through a conceptual pool, or swamp, depending upon their perspective! The life-jacket should provide buoyancy and provide its wearer with the confidence to continue on their journey. One of the aims of the book is to attempt to provide such a 'life-jacket' equipped with conceptual tools to analyse and perhaps give evocative explanations of the social world; second, the central aim is to convey the importance and significance of social theory for the disciplined study of society and social relations.

C. Wright Mills' in *The Sociological Imagination* (1959) powerfully narrates that the social analyst who is armed with 'thinking tools' seeks to understand the relationship of 'personal troubles', 'public issues', and the intersection of biography and history. These polarities, and the ways in which social analysts address them, define social theory. Mills further explicates that the role of the social analysts is to reveal the connections between what is going on in the world and what may be happening within ourselves - in other words, 'to grasp history and biography and the relations between the two within society' (1959: 33). The ability to shift perspectives, to analyze an experience or an issue from many levels of analysis, to see the intersection of these levels and mutual influence is at the very heart of the 'sociological imagination'.

Mills believed that social analyst develop a quality of mind that enables them to show individuals how their own 'private troubles' may be linked to features of the 'public world' of everyday life. Social theories provide analytical frameworks for understanding the interplay between human lives and changing social structures; theories are interested, therefore, in how society works, the arrangements of its structure and institutions and the mechanisms of its processes of social change.

There are three features of social theory which render its study difficult. First, all theories are not simply neutral mediums of interpretation. They contain a 'persuasive' power which seeks to render their accounts either superior over, or different from, those of others. Second, as will become apparent, social theory not only seeks to account for the dynamics of society, but also reflects upon how knowledge about the social world is possible, as well as the nature of 'social reality' (questions of epistemology and ontology, respectively). Third, social theories vary according to the cultural and intellectual traditions under which they are conceived and from which they draw their inspiration.

As a result of these characteristics, its study can appear to be inward-looking, frustrating, narrow, inapplicable and its ideas incommensurable and unduly 'philosophical'. Nevertheless, to theorize is a practice which either reflects upon, or assumes, the nature of knowledge and ideas of human action and consciousness. This means that there is a fine line between philosophy and social theory; which, it should be added, is a characteristic of sciences in general (Williams and May 1996). Therefore, in order to situate social theory, an understanding of the history and relationship between philosophy and social theory, as well as its cultural antecedents, is required. The other important point to note is that theories often mirror the norms and values of their creators and their social times, reflecting culturally dominant views of what should be the appropriate way to analyse social phenomena (Ritzer 2004). As much as important to understand the assumptions of social theories it is also necessary to understand the implications for understanding the social world.

Chapter 2

ENLIGHTENMENT AND SOCIAL THEORY

INTRODUCTION

Together with the intellectual legacies considered so far, the traditions of thought examined in this chapter had profound implications for the future direction of social theory.

This is not an easy history to document and for this reason the chapter is divided into two main sections. In the first section of the chapter we shall examine four traditions of thought: hermeneutics, phenomenology, pragmatism and language. For ease of understanding the first two are examined in one section.

Although there are clear differences between and within these four traditions, they all contributed to a shift in theoretical focus towards understanding the role of history, human action, meaning and language in everyday life.

In the second section four traditions of social thought are then considered: critical, structuralist, systems and feminist theories. Each of these have also added to and informed the direction of modern social theory. In order to specify how these traditions are connected to ideas in Part II of the book, each section will conclude with a summary discussion.

THE ENLIGHTENMENT: MODIFICATION AND CRITIQUE IN THREE TRADITIONS

Hermeneutics and Phenomenology: History, Consciousness and Being

Although there were differences amongst the neo-Kantians, they did share a scepticism regarding the possibility of utilizing a model of the natural sciences for the development of a 'science of society'. In so doing, they placed Enlightenment 'scientism' in question:

> 'Scientism' means science's belief in itself: that is, the conviction that we can no longer understand science as one form of possible knowledge, but rather must identify knowledge with science.
>
> (Habermas 1989a: 4)

The challenge to scientism, in terms of a separation between the realm of nature and human activity, informed the writings of Dilthey. This, however, leaves two questions. First, if the social sciences could no longer march in the shadow of the natural sciences, what was the alternative? Second, what implications does this have for the direction of social theory?

Dilthey's answer to the first question was to turn to an assessment of the scientific validity of historical writings. In the process, he sought nothing less than to provide new epistemological foundations for the human sciences. In this attempt, the influences so characteristic of German thought at this time emerged in his ideas:

> Kantian memories, strict historical training, a nostalgia for the spiritual world of the Enlightenment, a respect for the aims of positivist investigation, and, still hovering in the background, the misty pan-theism of Romantic theology.
>
> (S. H. Hughes 1979: 193)

This latter influence came from Schleiermacher (1768-1834) in whose writings the seeds of hermeneutics, defined as the theory and practice of interpretation, were laid. The question was how to produce a scientifically based hermeneutics? For this purpose texts had to be situated and interpreted in the social context in which they were produced. It would then be possible to replicate a central characteristic of science: to move beyond 'particularity' to the 'general'. To achieve that:

> it is necessary to rise above, not only the particularity of texts, but also the particularity of the rules and recipes into which the art of understanding is dispersed. Hermeneutics was born with the attempt to raise exegesis and philology to the level of a Kunstlehre, that is, a 'technology' which is not restricted to a mere collection of unconnected operations.
>
> (Ricoeur 1982: 45, original italics)

From a hermeneutical point of view, the social analyst must then start with a text and seek to understand it through an act of interpretation which connects them with the author and what was known by them when they constructed the text. What is often referred to as the circle involves a relationship between the part and the whole or, to put it another way: 'the relation between an isolated expression or work and the pre-given totality of language or literature' (Thompson 1981: 37).

This move also paved the way for a general inquiry into the nature of human understanding. Therefore, in seeking to fulfil his aims, Dilthey moved from the perspective of individualistic psychology, to an historicist view based on Hegel's objective 'mind' or 'spirit' as indicative of a given age:

> Every single human expression represents something which is common to many and therefore part of the realm of objective mind . . . the individual always experiences, thinks, acts, and also understands, in this common sphere.
>
> (Dilthey quoted in Outhwaite 1975: 26-27)

Dilthey could now argue that the human sciences could respect both individuality and universality through their ability to locate the 'inner' experiences of people in terms of their history. This contrasts with positivism, whose scientific aim is to explain the 'outer' observable patterns of human behaviour in terms of the relations of cause and effect. Nevertheless, how are these 'understandings' to be validated?

A solution to this problem was the adoption of verstehen (interpretative understanding). This allows the social sciences to possess something in addition to the natural sciences through their ability to understand the meaning of an event for social actors. After all, the natural sciences explain an event by considering it as 'the effect of a cause' (Strasser 1985: 2, original italics). To study human relations, on the other hand, must involve 'understanding what makes someone tick' or how they 'feel or act as a human being' (Taylor 1981: 30). A precondition of theoretical development then becomes an understanding of the consciousness which informs human actions.

Adding a voice to this breakdown of the unity of the human and natural sciences was the philosopher Edmund Husserl (1859-1938) whose concerns were, once again, a defence of the 'legitimacy and dignity of the human sciences' (Nuyen 1990). Husserl's route was to seek to establish 'truth' in the realm of human consciousness. This led him to a concern with perception in the attempt to grasp something called 'true understanding'. The method he advocated for this purpose was to question, or suspend, any presuppositions regarding society by adopting the strategy of 'bracketing'. People would then be freed from their historical context and enabled to uncover a 'true understanding' which is pre-theoretical and free from prejudice. Given this:

> Phenomenology asks us not to take received notions for granted. It asks us to question them - to question nothing less than our culture, that is, our way of looking at and being in the world in which we have been brought up.
>
> (Wolff 1979: 500, original italics)

Without bracketing experiences in this way the result would be, in what Husserl called the 'natural attitude' of everyday life, people simply taking-for-granted the world around them.

Husserl's teacher, Franz Brentano (1838-1916) stressed the idea of 'intentionality' in human psychology whereby consciousness is always directed towards 'something'. Husserl then took Descartes act of perception and added an important dimension: the 'thing' which is perceived. As such, objects of thought, including society and social relations, are viewed as being constituted in and through consciousness. In contrast to positivism, there is no separation of the social world from the act of perception, for it is an object of 'consciousness'. Husserl thereby transforms Descartes famous statement into ego cogito cogitata, translated as 'I think, and what I am thinking are the objects of my thoughts' (Strasser 1985: 124)

Knowing' is now reduced to a perceiving subject. For this reason, a phenomenologically informed social theory is required to focus upon a description and understanding of subjective perceptions, not an explanation of objective conditions. Questions such as 'What is social reality?' are then replaced by two different questions. First, 'How does reality come to be constituted by mental operations as a known object?' Second, 'How do we go about constructing our ideas of what reality is?' (T. Johnson, Dandeker and Ashworth 1984: 78)

Although discussions of phenomenology are found under the broad heading of 'interpretivist social theory', there are important variations in its unfolding history. After all, in their different ways, Husserl and Dilthey were still pursuing an Enlightenment ideal: that is, the search for 'true' knowledge. Their approach was to be challenged and modified in the writings of Martin Heidegger (1889-1976) who moved the focus of phenomenology from epistemology to ontology. In the process he was critical of Kant. He argued that we are not just observers of an external world mediated through our consciousness (Husserl), but are members of a world who exist as 'beings-in-time'. This moves away from Dilthey's neo-Kantian preoccupations with method, to the study of what Heidegger called Dasein (Williams and May 1996). This is best described as 'pre-understanding':

> the place where the question of being arises, the place of manifestation; the centrality of Dasein is simply that of a being which understands being.
>
> (Ricoeur 1982: 54, original italics)

This is not an easy notion to grasp. For our purposes, however, what Heidegger is criticizing is the concept of a 'transcendental ego'. He does not attempt to 'solve' the subject-object dichotomy by prioritizing human consciousness, as had Husserl and Dilthey, because understanding is seen as part of a 'mode of being'. It emerges, for example, from the gap which exists between where people are located in time and the possibilities which are available to them for the future. These possibilities are then 're-projected' into the present which can become 'an object of potential questioning, challenge, and change' (Bauman 1978: 166). The implication being that social theory should cast its gaze upon historically given subjects who exist in a 'lifeworld' in which a series of possibilities for social action are then contained. The practices within the lifeworld contain interpretations of, as well as methods of coping with, social relations. Hermeneutics is now the act of interpreting people's everyday interpretations.

Making Connections with Part II

Hans Georg Gadamer (1975) is largely responsible for the contemporary interest in hermeneutics. Following Heidegger, Gadamer does not see hermeneutics as a method for study, but part of the condition of human existence. These arguments have had an effect on the development of the social sciences. For instance, Jurgen Habermas seeks to develop a 'depth hermeneutics' which draws upon but is also critical of Gadamer. Heidegger's ideas appear in the work of Jacques Derrida, Michel Foucault and Norbert Elias (1897-1990) whose 'process' theory represents a critique of the subject-object dualism in social thought (see Elias 1991 ; Mennell 1992). Additionally, structuration theory will be shown to be ontologically based in terms of an emphasis upon human 'being' and 'doing', whilst Pierre Bourdieu considers the sociological implications of Heidegger's philosophy for generating a theory of human practice.

A more general phenomenological input into social theory is apparent in the work of Alfred Schutz (1899-1959). Schutz was a central figure in the development of phenomenological sociology, from which ontology derives part of its inspiration. Schutz rejected positivism as a model for the human sciences and was critical of the concept of verstehen. He did not view this as a method for 'doing' social science, but social science should study how people interpret the social world. In other words, how it is that people produce social life in a meaningful way. As Berger and Luckmann put it:

> Commonsense contains innumerable pre-and quasi-scientific interpretations about everyday reality, which it takes for granted. If we are to describe the reality of commonsense we must refer to those interpretations, just as we must take account of its taken-for-granted character - but we do so within phenomenological brackets. (Berger and Luckmann 1967: 20)

Together with Winch, who imported the insights of the philosopher Ludwig Wittgenstein (1889-1951) into social science, Schutz added to the focus of social theory on the rule-following and meaningful nature of human actions in everyday life.

The Pragmatists: Action, Consciousness and the Self

Little in the way of American social thought was noted in chapter 1, aside, that is, from a reference to Veblen. North America was left to be 'imported' into European social theory via Alexis de Tocqueville's (1805-59) interests in democracy. All of this was to change in the nineteenth century with the work of those such as William Sumner (1840-1910) and Edward Ross (1866-1951). Nevertheless, this did not involve travelling down a Hegelian-historicist road; one reason being that they regarded their own revolution, in contrast to that of the French, as a success (Ross 1991). This was to change with events such as the American Civil War and the process of industrialization.

Towards the end of his life Durkheim devoted a set of lectures to pragmatism for two main reasons. First, he saw in pragmatism an attack on rationalism which had informed so much of French intellectual life and was thus a threat to French culture in general (Joas 1993). Second, he viewed pragmatism as undermining the idea that 'true' knowledge was possible (Durkheim 1983: 1). Contained with these lectures was a veiled attack on his contemporary Georges Sorel, who distrusted intellectualist formulations and insisted: 'on the origin in practical action of all valid social theory' (S. H. Hughes 1979: 176, original italics). Given this we must ask what are the main elements of pragmatism which provoked such reactions? One approach to this question allows us to compare pragmatism with German-idealist attempts to ground epistemology in human consciousness:

> the basic idea of pragmatism, namely that it is actions rather than consciousness which are the foundations of thought, was developed in the 1870s by a group of young thinkers in Cambridge, Massachusetts, and was first publicly voiced by Charles Peirce in 1878.
>
> (Joas 1993: 95)

Charles Peirce (1839-1914) began his studies with the prejudices that people actually possess in everyday life, not with Descartes abstract method of 'doubting'. The Cartesian solitary ego then became replaced by a 'cooperative search for truth for the purpose of coping with real problems encountered in the course of action' (Joas 1993: 19). A similar argument also evolved in the work of John Dewey (1859-1952). For Dewey, ideas and actions should be judged according to the social situations which give rise to them. In contrast to the 'top-down' approach of metaphysical speculation, the social sciences should now concern themselves with the concrete social conditions under which actions take place. To add to this, William James sought to critique the idea of consciousness as a resolution to the object dilemma:

> Consciousness, he says, 'is the name of a nonentity, and has no right to a place among first principles' . . . he is not denying that our thoughts perform a function which is that of knowing ... What he is denying might be put crudely as the view that consciousness is a 'thing'.
>
> (Russell 1955: 840)

The consequences of these positions are far-reaching. Pragmatism displaces the relationship between thought and reality in terms of how truthfully one may be represented by the other. Thinking is now seen as a social, not solitary act, which takes place within a 'community of others'. Over time, however, this insight became distorted such that knowledge was to be judged in terms of its 'use-value'. The implication being that social practice is the means through which social theory becomes judged and validated.

Here we may detect a link between pragmatism and Marx's concept of praxis. That said, Marx would not have accepted the 'instrumentalism' of this position such that: 'knowledge should be judged as more or less "useful" rather than as true or false' (A. Sayer 1992: 70). The parallels with Heidegger's reflections, on the other hand, are more readily apparent:

> At the center of the first section of Being and Time stands the analysis of the concept of world. From the perspective of manipulation, and in general of nonobjectifying, practical dealings with physical com-ponents of the lifeworld, Heidegger explicates a concept of world as a network of involvements reminiscent of Pragmatism.
>
> (Habermas 1992a: 148, original italics)

Heidegger and the pragmatists may be seen to be calling upon social theory to shift its attention away from consciousness towards people's adaptions and methods of coping with the world in their everyday lives.

Making Connections with Part II

The clearest way in which the influence of pragmatism may be traced in social theory is via the work of George Herbert Mead (1863-1931) the Chicago School of sociology and the symbolic interactionists. These are the subjects of chapter 3. Mead's work, in particular, represents the desire to overcome the subject-object dualism in social thought. To recall the discussion so far. Following Kant, the individual had been considered as both 'a generating and autonomously acting subject' (Habermas 1992b: 158, original italics). As Mead goes on to suggest, the conceptualization of the subject then 'fell between the cracks' of a transcendental ego, apparently capable of surveying the world around it through the exercise of reason and an empirical ego, who simply exists as one among many in the social world.

In Mead's work we will find neither an interest in utility-oriented action (towards a desired end) or a Kantian style moral-individual action (transcendental ego), but an examination of the relationship between consciousness and action. Social action is not conceptualized in terms of an orientation to a predetermined end, for the very consciousness which formulates an individual's ends cannot take place outside of the context of the action itself:

> Rather, the setting of an end can only be a result of reflection on resistances encountered by the variously orientated behaviour of a life form whose world is always already schematized in a practical manner prior to all reflection.
>
> (Joas 1993: 248)

Social 'selves' are viewed as being formed through a process of confronting the world of which the individual is a part in dialectic relation between subject and object. Only by encountering objective conditions are people able to gain a sense of 'who' they are.

There are a further six ways in which connections between pragmatism and modern social theory can be drawn. First, in Peirce's work on semiology (the study of signs and their systems) the idea of 'indexicality' appeared. This suggests that we may understand an utterance only in the context in which it is made. As we shall see, this is part of the theoretical armoury of ethnomethodology. Second, there is Habermas's theory of communicative action. Whilst his critical theory has foundationalist aspirations, the linguistic aspects of his work contain a pragmatist theory of meaning which revolves around the assertion 'that to understand an utterance is to understand the claim it raises' (Cooke 1994: 95). Third, pragmatism forms an essential part of the ideas of the contemporary 'anti-foundationalist' American philosopher Richard Rorty (1982). Despite his own reservations on this trend in social thought (Rorty 1993) his works have added to the 'postmodern turn' in social theory. Fourth, Alfred Schutz used James' ideas in order to develop his phenomenological social theory. Fifth, pragmatism existed in the critical thought of Charles Wright Mills (1916-62) who called for a fusion of the practical and intellectual within social science (Mills 1970). Finally, alongside the ideas of Durkheim, phenomenology, linguistics and ethology, we shall uncover its influence in the works of Erving Goffman (1922-82).

Having examined these philosophical schools in terms of their implications for the development of social theory, attention is now turned to four traditions of social thought which represent the journey of renewal, modification and critique in the Enlightenment project.

Chapter 3

FOUR TRADITIONS IN SOCIAL THOUGHT

CRITICAL THEORY: FOR (AND AGAINST) MODERNITY

The term critical theory has been used in a plethora of contexts with different meanings but always probes beneath the everyday surface of social reality to find what might be hidden there. It has been used to denote a mode of enquiry into culture and history in order to access deeper levels of meaning and understanding. Within the context of social theory, it has been used to describe an:

> 'attempt to grapple seriously with the historical embeddedness of all theory, where a full understanding demands an appreciation of the historical and cultural conditions ... on which the theorist's own intellectual activity depends.
>
> (Calhoun 2000: 537)

In a specific sense, it has been used to describe the work of the Frankfurt school of social theorists who were to attempt to fuse Marx and the work of Freud (Layder 2006). Critical theory has two inter-related aims: to bring to consciousness, an awareness of capitalist exploitation, bureaucratic domination and to create a popular demand for liberation. Critical theory has two main aims: it brings to our consciousness oppression of which we may or may not have been aware, and it calls for "criticism of life" to resist and change the existing system of domination and exploitation (Powell and Moody 2003: 11).

The fundamental aim of critical theory is the reconstruction of truth and our existing society of oppression (Habermas 1992): racism, genocide, increasing environmental pollution, and possible nuclear holocaust cannot be the "truth" of human existence. Truth must lie somewhere else, not in the facts of the given reality, but in the negation or transcendence of those facts. Truth lies in the attempt to go beyond this reality to a better world. Thus, truth lies in our attempt to change the world, in a critique of established reality. A critical theoretical analysis is true insofar as it helps change the world and make it a more human place in which to live. So while reality may lie quite beyond human comprehension and can never be reduced to words, truth is the living of each moment of human life to its maximum potential, the unfolding of reality from its potential to a lived practice or actuality of human existence. In contrast to critical theory, much of sociology remains at the level of the "facticity of the given," (Marcuse 1964: 170). Such analysis has the effect of justifying the status quo whether it intends to do so or not.

Coupled with this, is the notion that the truth of a theory is a *praxis*, or the degree to which sociological analysis is responsive to human values (Habermas 1992). Praxis is the key concept that differentiates the critical sociologist from the ahistorical gatherer of "common sense facts" and cataloguer of mathematical abstractions whose activities characterise positivism. Praxis refers to the ideal of conscious practical action - of making the critique of alienation speak for popular needs and lead to concrete actions against the capitalist commodity relationship within historical possibilities. Knowledge serves "real world" interests such as statistical knowledge in the case of sociological positivism which assumes a special status from which to view social reality, status that claims to be representing the truth about reality. No matter what pains are taken for "objective value-free analysis," (Weber 1949) knowledge cannot be divorced from social reality and values.

The historical foundations of critical theory are in the early philosophical writings of Karl Marx. Prior to the 1840's, social theory, political theory, "scientific" sociology, and classical political economy existed as coherent intellectual traditions, but a critical theory of society had not yet emerged. Utopian socialism, scientific political economy, and critical philosophy existed, but were not yet combined in a critical theory of society which could reflect the class struggles and social revolutions which were transforming Europe in the middle of the 19th century.

Critical theory became bound up with a series of pathways for intellectual inquiry that first emerged with the end of the 18th century European Enlightenment and in particular with the initial widespread waning of intellectual confidence that the newly hegemonic bourgeois society would succeed in realizing Enlightenment ideals. In short, critical theory represents the intellectual articulation of the conviction that modern capitalist society cannot - at least not without significant reformation or substantial transformation - realize the Enlightenment ideal of a rational, just, and humane society. According to Enlightenment consensus, this (ideal) society is to be one which will genuinely embody the highest values of human civilisation, and which will thereby insure steady progress in the attainment of liberty, justice, prosperity, and contentment for all of its citizens.

Social theory in the work of Hegel, political theory in Rousseau (1950), and positivist and "scientific" sociology in Comte (1966) and Spencer (1969) were either grounded in a criticism of the human condition in political society, or in faith in the possibility of a natural science of the enduring laws of society. Criticism of society was reduced by these theorists to criticism of the state, and critical thought was reduced to a search for formal laws of social structure and development. It was not until Marx that the basis for a truly critical and social theory was established. In his early work, Marx sought release of humanity from the bonds of alienation from the domination by a society of wage labour and economic exploitation.

Critical theory was further influenced by George Lukacs (1881-1971) in his *History and Class Consciousness*. Lukacs was appalled by the barbarism and depravity of World War I and inspired by the Russian Revolutions of 1917 and the Hungarian Commune of 1919. At the time that he finished Marx's economic and philosophical manuscripts were gathering dust in Moscow, unpublished and unnoticed for nearly seven decades.

Under the auspices of its second director, Max Horkheimer (1895-1973), the Institute for Social Research at Frankfurt came to centre its interests around six broad areas. First, the seeking of explanations for the absence of a unified working-class movement in Europe. Second, an examination of the nature and consequences of capitalist crises. Third, a consideration of the relation-ship which existed between the political and the economic spheres in modern societies. Fourth, to provide an account for the rise of Fascism and Nazism as political movements. Fifth, to study socialization in families in terms of ego development. Finally, to undertake a sustained critique of the link between positivism and science (adapted from Held 1990: 35).

In order to adequately examine these questions a new theoretical approach was required which involved a 'refusal to champion vulgar idealism as an antidote to vulgar materialism' (Jay 1973: 294). This involved a return to the Hegelian strand in Marxism in order that the concept of human praxis might be recovered from overly materialist readings of Marx. Yet between Hegel and the Frankfurt School stood such figures as Schopenhauer and Nietzsche, both of whom had questioned the Enlightenment project. Therefore, it was necessary to engage with these ideas in order to recover the promise of Marxism.

To these intellectual antecedents and aims we must add the political context in which these scholars wrote in order to fully situate their work. In Nazi Germany, the Institute found itself under threat with the result that its leading members, including Theodor Adorno (1903-69), Max Horkheimer and Herbert Marcuse (1898-1979) emigrated to the United States (Adorno and Horkheimer returned after the Second World War). Here they found the self-confidence of bourgeois liberal-capitalism with its apparent ability to absorb and neutralize proletarian consciousness. To analyse how this occurred necessitated a consideration of culture which, up to this point in time, had been largely devalued in Marxist circles (see Raymond Williams 1961).

The work of these scholars was inter-disciplinary, accompanied, as noted, by a rejection of 'scientific Marxism', together with critiques of positivism and interpretivism. As Horkheimer said of the relationship between social philosophy and positivism:

> Contemporary social philosophy, as we have seen, is in the main polemically disposed toward positivism. If the latter sees only the particular, in the realm of society it sees only the individual and the relations between individuals; for positivism, everything is exhausted in mere facts. These facts, demonstrable with the means of analytic science, are not questioned by philosophy.
>
> (Horkheimer 1993: 7)

Positivism turns human relationships into nothing more than abstract categories through a failure to examine the conditions under which they develop and are sustained. In this sense it represents part of the increasing desire to control the social and natural worlds in the name of profit (Adorno *et al.* 1976). What makes critical theory 'critical' is precisely this recognition of the connection which exists between knowledge and human interests.

Interpretivist social theories were also inadequate due to an exclusive concentration upon the process of self-understanding and self-consciousness. The result being that they exhibit an uncritical acceptance of dominant forms of consciousness within given societies and fail to consider the structural determinants of human actions, as well as the role of the unconscious and power and control in human relations. Indeed, in another direction, critiques of this type led to attempts to fuse phenomenology and Marxism (Paci 1972) and an emphasis upon ideological mechanisms of control as apparent in the works of Georg Lukacs (1885-1971) and Antonio Gramsci (1891-1937). It was Gramsci who employed the concept of hegemony whereby it is held that the bourgeoisie exercises its social control through both coercion and consensus within institutions. What presents itself as common sense is actually symptomatic of ideological distortion (Gramsci 1971). This notion still inspires cultural studies (Nelson and Grossberg 1988) as well as leading to a framework for 'post-Marxist' politics (Laclau and Mouffe 1990).

Beginning in Germany, a tradition has extended and enriched the critical theory of society which includes the persons associated with the Frankfurt Institute for Social Research - Marcuse, Adorno, Horkheimer, Benjamin and Habermas, among others (Jay 1973). The rise to power of Hitler in Germany nearly destroyed the Frankfurt School.

Critical theory progressed from the early Frankfurt school to researchers who now attempt to become actively engaged in promoting social change within the education system and the culture itself. They seek to promote change by 'becoming part of the self-consciousness of oppressed social groups' (Hoy and McCarthy 1994). The Frankfurt School grew out of the Institute of Social Research, which was founded in 1923 at the University of Frankfurt by Felix Weil, a political scientist with a passion for Marxism. One of the major purposes of the institute was to study (and eventually explain) the dynamics of social change. Some versions of Marxism are critical theory especially as regards Frankfurt School but in recent years has been applied to other social philosophies of linguistics, psychoanalysis and existentialism.

The 'critical theorists' themselves focused upon the concept of ideological incorporation. Herbert Marcuse (1968a) spoke of the 'technical-administrative control' of society and took aim at a growing corpus of social theory which claimed to have witnessed a vindication of capitalism through what was known as the 'end of ideology' thesis (Bell 1960):

> The capabilities (intellectual and material) of contemporary society are immeasurably greater than ever before -which means that the scope of society's domination over the individual is immeasurably greater than ever before. Our society distinguishes itself by conquering the centrifugal social forces with Technology rather than Terror, on the dual basis of an overwhelming efficiency and an increasing standard of living.
>
> (Marcuse 1968a: 9)

It follows that the rationality of 'the system' itself had to be subjected to critique. One way in which this took place was through an investigation into the Enlightenment project. Here we find history interpreted as a journey towards a 'new kind of barbarism' whereby:

> The individual is wholly devalued in relation to the economic powers, which at the same time press the control of society over nature to hitherto unsuspected heights . . . The flood of detailed information and candy-floss entertainment simultaneously instructs and stultifies mankind.
>
> (Adorno and Horkheimer 1979: xiv-xv)

On this basis Weber was criticized for confusing capitalism as a process of rationalization with reason itself (Marcuse 1968b). Instead of liberating the subject, the Enlightenment had witnessed a growing disjuncture between reason and rationality: 'The unleashed market economy was both the actual form of reason and the power which destroyed reason' (Adorno and Horkheimer 1979: 90). The result being a 'fractured' subject through self-alienation perpetuated by positivism with its fetish for collecting social facts. Modernity now represents a loss of meaning:

> For the Enlightenment, whatever does not conform to the rule of computation and utility is suspect. So long as it can develop undisturbed by any outward repression, there is no holding it . . . Every spiritual resistance it encounters serves merely to increase its strength . . . Enlightenment is totalitarian.
>
> (Adorno and Horkheimer 1979: 6)

This meta-critique noted, the question remains that having rejected interpretivist social theory how can the important analysis of mental states be achieved? This is where Sigmund Freud enters as part of the whole history of the relationship between psychoanalysis and social theory (see Elliott 1992).

Freud sought to supplant the idea of the conscious process of human thought with a theory of the unconscious. Whilst he acknowledged that an unconscious process could be inferred only from observing patients, he believed it to play a dominant role in the determination of behaviour. He thus placed the idea of the 'conscious subject' into question, but for different reasons that any we have noted so far:

> According to Freud . . . we are not who or what we think we are: we do not know our own centers; in fact, we probably do not have a center at all. Psychoanalysis undermines notions about autonomy, individual choice, will, responsibility, and rationality, showing that we do not control our own lives in the most fundamental sense.
>
> (Chodorow 1986: 88)

This focus moves away from an understanding of the solitary ego in the social world, to the self-misunderstanding person who fails to see the causes of their own symptoms. Indeed, Freud's whole metapsychological project can be read as one which sought to show 'who we are', whilst psychoanalysis is a practice for the development of autonomy and control in the 'self' (Chodorow 1986).

More specifically, Freud viewed all behaviour as being motivated or caused whether it be 'neurotic' or 'normal'. This could be attributed to earlier determinants which appeared as symptoms whose purpose was to avoid painful memories. In the self there exists a continuous struggle between 'internality' and 'externality' or, to express it another way, between mental structures (id, ego and superego) and life-experiences. This, however, does not simply take place 'within' the individual, but is a response to their physical and social environments. These may be manifest in a whole series of 'slips' (Freud 1938) and the idea of civilization itself (Freud 1969).

A Freudian approach to the unconscious and self-misunderstanding permitted the Frankfurt theorists to concentrate on the process of ideological distortion under capitalism, without resorting to some notion of consciousness as an explanatory framework: 'it is only through the process of interpretation that subjects can come to know the conditions and complexities of their own existence' (Thompson 1981: 212).

Despite this utilization of Freud, we are still left with problems. Up to the time of the Frankfurt theorists Marxists had dismissed psychoanalysis as unworthy of attention (see Osborne 1937). After all, the problematic fusion of Marx and Freud is apparent at different levels:

> On the practical level, psychoanalysis is carried on by rich doctors on richer patients. Conceptually, it starts from the individual's problems and tends to play down social conditions and constraints.
>
> (Ollman 1979: 177)

If Freud's work was to serve the needs of critical theory, it has required modification. At this point, Leo Lowenthal and Eric Fromm enter the scene as central to this process. Fromm, for example, examined the relationship between the individual and totalitarianism (Fromm 1969). He also held throughout his life that our capacity for love and freedom: 'depends almost entirely on the given socio-economic conditions' (quoted in Jay 1973: 100). The interest in Freud was also apparent in socio-psychological studies on the structure of modern personality types (Adorno *et al. 1950)* as well as in the writings of Wilhelm Reich on the relationship between sexuality and capitalism (see Ollman 1979). Fromm also examined politics from a psychoanalytic point of view:

> It has been shown that this irrational acting and feeling are the result of certain instinctual impulses, of which the actor is often unconscious, but which compulsively condition him or her. It therefore seems to follow that psychoanalysis could also provide the key to an understanding of similarly conditioned social action, of similarly irrational political events.
>
> (Fromm in Bronner and Kellner 1989: 213)

Making Connections with Part II

The critical theorists left a legacy which is still the subject of new interpretations and understandings (Bronner 1994; Hoy and McCarthy 1994; Wiggershaus 1995). Of all contemporary scholars, it is Jurgen Habermas who sits most firmly within this tradition but who, as we shall see, has submitted it to extensive modification. However, if we broaden our horizons to trace a critical 'intent' in the development of social theory, the picture becomes less restrictive. For instance, Pierre Bourdieu sits within a tradition which runs from Marx to the Frankfurt School (Calhoun 1995). There are also the ideas of Michel Foucault, whose own words suffice here:

> As far as I'm concerned, I think that the Frankfurt School set problems that are still being worked on. Among others, the effects of power that are connected to a rationality that has been historically and geographically defined in the West, starting from the sixteenth century on. The West could never have attained the economic and cultural effects that are unique to it without the exercise of that specific form of rationality.
>
> (Foucault 1991a: 117)

Whilst the Frankfurt scholars were writing, another tradition was active in French intellectual life. Starting with an examination of language and working through to anthropology, it culminated in the works of a Marxist thinker who, unlike the above, sought to expunge any Hegelian trace in Marxism to provide it with a 'scientific' status.

STRUCTURALISM: FAREWELL TO THE SUBJECT?

The French intellectual climate was undergoing rapid change in the post-Second World War period. Existentialism and phenomenology were vying for intellectual ascendancy with mixtures of Marxism, the psychoanalysis of Freud and Jacques Lacan (1901-83) and interpretations of the structuralist writer Ferdinand de Saussure (1857-1913).

Structuralism may be distinguished from other traditions by noting its contention that knowledge cannot be grounded in individuals nor their historically given situations. Conceptually, structuralism asserts that the structure of discourse itself produces "reality" - that we can think only through language, and therefore our perceptions of reality are all framed by and determined by the structure of language. Further, that discourse dominates individuals; that the source of meaning is not an individual's experience, but the sets of discourses that govern society (Powell and Moody 2003).

Structuralism also rejects empiricism for its failure to distinguish between the appearance of a phenomena and its underlying 'reality'. It is a 'science of relationships' with the emphasis of holistic structuralism being upon the interconnections or interdependencies between different social relations within society (Runciman 1973). This position holds that we cannot understand the dynamics of a society by studying each of its parts in isolation. Instead, we can explain its parts only through their relationship with other constituent elements in a given system. To give an everyday example. We stop at a set of traffic lights because we contrast the signs with each other. Red signifies that we should 'stop' because it is contrasted with green which means 'go': 'When we see red we see red-and-not-green' (Keat and Urry 1975 :124).

Saussure, the founder of modern structural linguistics, separated the regularities or codes of language (la langue) from its usage by an individual who externalizes the underlying system (la parole). Saussure argued that there is no consciousness before language. He considered the relationship between thought and language via a comparison with a sheet of paper:

> thought is the front and the sound the back; one cannot cut the front without cutting the back at the same time; likewise in language, one can neither divide sound from thought nor thought from sound . . .
>
> (Saussure in Easthope and McGowan 1992: 7)

Language itself is grasped through an ability to assimilate and order a system of signs which are its basic units. Signs are divided into two parts: the signifier (the sound) and the signified (the idea or concept to which the sound refers). The latter does not refer to an actual object in the material world, but the way in which the system of language, through its signs, organizes the world, with signs becoming defined by their differences within the language system itself. In this way, the conscious individual is supplanted by a system of language which is prioritized over its manifest utterances in individual speech-acts. Given that the relationship of signs depends upon a network internal to the language system itself, they are seen to possess value before signification. Thus, within language, there are no absolutes, only relations of mutual dependence (as with our traffic light example):

> The idea or phonic substance that a sign contains is of less importance than the other signs that surround it. Proof of this is that the value of a term may be modified without either its meaning or its sound being affected, solely because a neighboring term has been modified.
>
> (Saussure in Easthope and McGowan 1992: 13)

As language is self-referential, it can be studied as an internal system through which both subjectivity and the social world are produced. Saussure envisaged that this may be applied to the non-linguistic realm as part of the science of semiology. It is the development of this insight that explains much of the history of twentieth-century structuralist thought. Consider, for example, observing a game of chess. It may be concluded that people adhere to common rules. An analysis of those rules would then indicate the strategies of the players, with a direct involvement in the game not being necessary for explanatory purposes. The same can be said of an analysis of food:

> The menu, for instance, illustrates very well this relationship between language and speech: any menu is concocted with reference to a structure (which is both national -or regional -and social); but this structure is filled differently according to the days and the users, just as a linguistic 'form' is

filled by free variation and combinations which a speaker needs for a particular message.

(Barthes in Lemert 1993: 343)

This focus upon language without reference to a speaking subject, uncovers an affinity with Freudian psychoanalysis. Whereas Freud prioritized the unconscious processes of human behaviour over consciousness, Saussure prioritized the system of language over the intentions and meanings for an individual speaker. Ricoeur also refers to this joint combination of attacks on the theory of the subject as the 'semiological challenge' (1982: 35).

Levi-Strauss (1963) starts his study of 'myths' with a dilemma: that is, their content is contingent but why, given this, are they so similar throughout the world? This is explained through the prioritization of form over content. In using the methods of Freud, he seeks to account for diverse social phenomena in the logic of the human mind itself. He argues that myths are not reducible to individual consciousness, but are themselves a product of a 'collective unconsciousness' (the influence of Durkheim being mediated here via Mauss). This exists prior to individuals and is constraining upon their actions and thoughts. Myths then work to organize social relations and they have a cognitive function (Stehr 2005).

In the study of myths Levi-Strauss argues that he has uncovered the universal structures of the human mind. These structures are derived from binary oppositions. The very ability of people to think is governed by dualisms: for example, between raw and cooked; us and them; ideal and material, etc. Whilst there may be cultural variations in the content of these oppositions, their underlying form is universal. Here we can detect a neo-Kantianism. Although not rooted in consciousness, verbal categories are considered as the mechanisms which link the universal structures of the human mind to the universal structures of culture. However, in this process the subject is displaced by structures beyond their conscious apprehension. This position is also found in the work of Louis Althusser (1918-90).

In contrast to Sartre's (1905-80) existential interpretation (1960) the structuralist reading of Marx enjoyed an enormous resurgence in Althusser's arguments (1969). His aim was to recover the scientific element of Marxism from the tradition of existentialism, without succumbing to the problems of a materialist philosophy of history. To achieve this he employed a 'symptomatic reading' of Marx (a term borrowed from Freud). Broadly speaking, he argued that there was an 'epistemological break' in writings -between the early (pre-

1845) and 'mature' Marx. The latter provided the scientific basis for an explanation of history thereby constituting a dismissal of the Hegelian legacy.

Whilst Engels had added to the one-sided materialist interpretations of Marx, he also wrote:

> According to the materialist conception of history, the ultimately determining element in history is the production and reproduction of real life. More than this neither Marx nor I have ever asserted. Hence if someone twists this into saying that the economic element is the only determining one, he transforms this proposition into a meaningless, abstract, senseless phrase.
>
> (Engels to Block in Marx and Engels1953: 498, original italics)

Althusser took this qualification as indicative of the 'relative autonomy' of the ideological and political spheres from that of the economic. How does this translate into structuralist language?

The answer to this question may be seen in the relationship between the system (society) and its constituent parts (people). Society pre-exists the individual and ideological practices provide the means through which people relate to the world around them. What Althusser calls 'Ideological State Apparatuses' that consist of the family, media, educational and religious institutions which in turn constitute who we are, how we think of ourselves and may even be the origin of our individual traits. In other words, they both work through us and yet enable us to recognize ourselves. We even learn the idea of the autonomous bourgeois subject from families and therapists. Speaking of which, Althusser used concepts from the writings of the psychoanalyst Jacques Lacan. Where Althusser sought to conceive of Marxism without Hegel's absolute subject, so thought of psychoanalysis without Freud's ego (Sarup 1993). Now, in the works of Althusser and his contemporary Nicos Poulantzas (1936-79), who sought to employ Althusserian concepts (Poulantzas 1978), structures do not come from the universal categories of the human mind (Levi-Strauss) or from language (Saussure), but from the history of class societies.

Making Connections with Part II

Whether through critique, revision or contrast, structuralism has generated extensive interest. In terms of contrasts, we can compare it with the utilitarian-based exchange theories of those such as Homans (1961) and the 'rational-choice' model of human behaviour which has enjoyed a well-argued defence in the work of James Coleman (1990). Revisions include the neo-structuralist Mary Douglas (1970; 1978) but her intellectual debt is more to Durkheim given an emphasis on ritual and symbol in social life.

By the mid 1970s Althusser's legacy had withered and the critiques were numerous. These included the French Marxist Henri Lefebvre (1901-91) and English historian E. P. Thompson (1924-93). Yet it can be argued that his legacy is apparent in the most productive area of contemporary Marxism: the analytic Marxism of Cohen, Elster and Erik Olin Wright, amongst others (see Mayer 1994). It is argued that the questions they pursue are only possible when Hegelianism is abandoned. In this sense, this school could be termed 'post-Althusserian Marxism' (Callinicos 2000).

As for critical contrast, the structuralist 'dissolving' of the subject has informed ideas on the social 'self' and the agency-structure debate. Anthony Giddens, for example, accepts the problems of prioritizing a 'philosophy of consciousness' in social theory, but does not accept the idea of a 'subject-less history'. As a thinker steeped in the traditions of French social thought, Pierre Bourdieu is clearly influenced by structuralism. Nevertheless, he aims to return agents, rather than subjects, back into the terrain of social theory. Structuralist connotations will also be found in Habermas's idea via his use of the work of the structural psychologist Jean Piaget (1896-1980) whilst Michel Foucault is often interpreted as working within this tradition.

Structuralism also appeals in the realms of literary criticism, cultural studies and history; the latter being exemplified in the work of the 'Annales' historians (see Clark 1991) who have influenced the 'world-systems analysis of Immanuel Wallerstein (1990). In terms of literary criticism, Jacques Derrida's (1930-2004) engagement with and critique of structuralism in terms of the arbitrary character of signs, has provided the catalyst for an enormous growth of interest in textual analysis (Kamuf 1991; Norris 1987). For one writer, Derrida has been responsible for starting an 'anti-humanism in the humanities' of almost epidemic proportions (Schwartz 1990). Finally, Jean Baudrillard treats commodities as systems of signs which have a reference only to each other.

Levi-Strauss developed his methods against the idea that society was essentially stable. He also prioritized structure over function. Yet the inspiration for normative functionalism, which we shall consider in examining the work of Talcott Parsons (1902-79) also stemmed from the structuralist writings of Comte and Durkheim. This found its century expression in the work of the social anthropologists Bronislaw Malinowski (1884-1942) and A. R. Radcliffe-Brown (1881-1955).

Systems Theory: Talcott Parsons and the Celebration of Modernity

The emphasis of functionalism, as with structuralism, tends towards the synchronic and holistic. On the other hand, normative functionalist explanations downplay the search for origins or causes of social phenomena in favour of seeing cultures as total systems of functionally interrelated parts. Teleological explanation is then employed whereby the statement: 'the function of X is to maintain Y, implies that X has some kind of causal influence on Y' (Dore 1973: 66). This may be illustrated through the Durkheimian argument that religion exists to provide the moral foundations of society. In order to see how this develops in one particular direction, attention will be focused on the writings of Talcott Parsons.

Parsons perceived a convergence in social and economic theory around the turn of the twentieth century. This centred upon a concern with the subjective experiences and actions of individuals combined with the Hobbesian problem of social order expressed in terms of the question: how are diverse interests in a society possible? Parsons then embarked upon the building of bridges between dualisms in social thought; in particular, between positivism, which ultimately regards human action as a passive adaption to the social world and utilitarianism, where only an actor's orientation to an end is required to understand their actions. According to this idea, social order results from the egoistic pursuit of ends from which is supposed to emerge the 'public good'. Parsons, as Durkheim before him, found this theory to be fundamentally flawed. After all, the pursuit of private ends would be best achieved by 'force and fraud' from which social chaos, not order, would emerge. He thus sought an explanation which moved away from the simplistic notion that action was motivated by self-interest. What form did this take?

Parsons's (1937) theory of action contained four parts. First, the actor. Second, the desired state of affairs towards which their action is oriented (this is the teleological aspect of the act). Third, the social situation in which the action takes place. Here, those social situations over which they have control are called 'means', whilst those over which they have no control, he calls 'conditions', including not only the physical environment, but also biological potential and limitations. Fourth, a 'normative orientation'. This normative orientation he regards as the 'motor' of social action (Parsons in Hamilton 1985: 73-78). Social action is thus conceptualized in terms of the use of means to reach desired ends in a process infused with societal norms. The Hobbesian problem is then resolved via a process in which the actor is seen to possess a sensitivity to the meanings in their environment which they then take account of their action:

> These normative rules both define what immediate ends should and should not be sought, and limit the choice of means to them in terms other than those of efficiency. Finally, they also define standards of socially acceptable effort. This system of rules, fundamental to any society . . . is what I call institutions. They are moral norms, not norms of efficiency.
>
> (Parsons in Hamilton 1985: 62, original italics)

In Parsons's 'second phase' (Hamilton 1983) in which he published *The Social System* (Parsons, 1951) the emphasis of his thought then shifted towards 'social structure', defined as:

> a system of patterned expectations of the behavior of individuals who occupy particular statuses in the social system . . . Such a system of patterned legitimate expectations is called by sociologists a system of roles. In so far as a cluster of such roles is of strategic significance to the social system, the complex of patterns which define expected behaviour in them may be referred to as an institution.
>
> (Parsons in Hamilton 1985: 126-127)

Analytic focus is now less upon the actor's rationality and more a specification of what is meant by 'normative orientation' in the choice of actions. Parsons is now more concerned with the means through which social organization may be explained in functionalist terms. Social stability is conceptualized in terms of social institutions which provide 'behavioural forces' which act as linking mechanisms between actions and systems. These institutions ensure 'that action can be sufficiently regularized so as to be compatible with the functional requirements of a society' (Parsons in Hamilton 1985: 127).

In these terms cultural patterns provide the framework for choices between courses of action. These options were reduced by Parsons to four dilemmas, each containing two responses with a definite meaning for the actor concerned. They are: affectivity versus affective neutrality; universalism versus particularism; achievement versus ascription and specificity versus diffuseness (a fifth, self-orientation versus collectivity-orientation, was later dropped). These may be manifest in the following way: the choice to set aside feelings or become more involved in a relationship; to judge an object according to general or specific criteria; to judge an object either in terms of its use, or in terms of a quality independent of its benefit and finally, to come into contact with others only in specific contexts or in many contexts, in which case relationships are more diffuse.

For Parsons the above could be used to categorize modes of action at increasing levels of abstraction: from the personality system, to the normative requirements of the social system and the value-patterns of the cultural system.

Within his overall convergence thesis, Parsons viewed Freud and Durkheim as working on the same problem. What was needed was a fusion of the internalization-externalization process which linked the needs-dispositions of personality with those of the social system. However, this did not mean that there exists a simple correspondence between the structure of personalities and social institutions because: 'persons occupying the same well-defined status in the social system will be found to cover a wide range of personality types' (Parsons in Hamilton 1985: 127). The need exists, therefore, for 'motivating mechanisms' which take account of such diversity.

Although his preoccupation with social integration is still apparent, following the influence of the physiologist Henderson, he now begins to see societies more as organisms. For Parsons all theories involved two elements: the static and the dynamic. The static aspect is, as with structuralism, the structure of the social system. Its dynamic elements are the means through which the static categories are linked; this is the concept of function. The relationship between subsystems of action and the social system as a whole is now of interest where functions provide the links between the dynamic and structural aspects of a social system. An emphasis upon cybernetics, defined as the study of communication between different elements in a social system, now becomes evident in his work.

Drawing on the work of Robert Bales and together with Edward Shils, Parsons then argued that each system, from the unit-act to the social system as a whole, had four functional problems. These were listed in order of significance as: pattern maintenance; goal attainment; adaption and integration (AGIL). Pattern-maintenance is concerned with socialization and its focus is upon the structural category of social values. Stability is the key here, involving 'tension-management' due to the potential of 'strain' between personality and the needs of the social system. Goal-attainment is accomplished by the state and polity for the purposes of establishing goals and the motivating elements in the system towards their achievement. Meeting the needs of the system in terms of the acquisition and distribution of resources is required to ensure equilibrium. In a situation of a plurality of goals, however, the question of their 'cost' relative to others will arise. Therefore, a need exists for the economic system to maximize: 'generalized disposability in the process of allocation between alternative uses' (Parsons 1965: 40). Lastly, as systems are differentiated and segmented into more or less independent units, there is the question of their adjustment and contribution to the effective functioning of the social system as a whole. Norms are then required to establish coherence and control between subsystems. As Parsons notes:

> In a highly differentiated society, the primary focus of the integrative function is found in its system of legal norms and the agencies associated with its management, notably the courts and the legal profession.
>
> (Parsons 1965: 40)

In sum, the functional maintenance of the social system is achieved through four functional prerequisites. For the purpose of social stability, values then link the need-dispositions of personality with that of the social system which, in turn, is linked via role-expectations to cultural consensus. Husserl's 'transcendental subject' is now replaced by a 'transcendental actor' given that Parsons sees society and culture as the transcendental conditions for human action (Bauman 1978).

By starting with meaningful action and then moving on to consider the cultural and social systems as coordinating mechanisms for that action, neither the individual nor society are prioritized.

Therefore, contrary to popular interpretations, he does not simply abandon the individual for the social system over the course of his work. It is more productive for the development of understanding social relations, to see his work as moving from an analysis of action to social interaction within social systems as a whole.

Making Connections with Part II

Perhaps the first point to note in reflecting upon Parsons's work was his optimism:

> industrial society is the end of a long road, the culmination of a slow process of maturation. It is the most finished form of society because social organization and human rationality find their finest expression within it.
>
> (Rocher 1974: 158)

We do not discover a longing for a bygone age of community in the face of the forward march of industrialism. On the contrary, the focus on values in modern society was undertaken without a nostalgia for the pre-modern. Yet Parsons was also aware of the major shifts within modern societies towards the expressive and symbolic. In that sense he anticipated some of the modern versus postmodern debates in contemporary social theory (B. S. Turner and Robertson 1991).

In terms of his more direct influences, Habermas, whilst subjecting his work to extensive critique, also utilizes the AGIL schema and systems' ideas (Habermas 1975). He writes: 'no theory of society can be taken seriously today if it does not at least situate itself with respect to Parsons' (1987: 199). Part of this is explained by Parsons's conceptual bridge building between continental and Anglo-American social theory (Sciulli 1985) and we find the systems theorist, Niklas Luhmann, interpreting Parsons in a way which owes more to Heidegger than Husserl. The social system is then seen to require increased mechanisms of control as society becomes more complex and existential anxiety is exacerbated. For the neo-fuctionalist Jeffrey Alexander, however, the system is then over-determined for, in Luhmann's conservative emphasis, no conflict is considered outside of the demands of the system (Alexander 2004). Therefore, we should note that for the more liberal-minded Parsons: 'culture is always a system analytically separated from society' (Alexander 1984: 402).

In the United States Robert Merton (1968) was to modify the functionalist tradition. In particular, his preference for a 'middle-range theory', more amenable to empirical testing, was in tune with American intellectual culture (Bershady 1991). Further, one of Parsons's graduate students, Harold Garfinkel, went on to consider the gaps in his action theory and became the founder of ethnomethodology. Whether in terms of continental or Anglo-

American social theory, or as a theorist against which to sharpen theoretical axes, his legacy cannot be ignored.

Embodied in Homans's (1973) article, 'Bringing Men Back In', is the criticism that Parsons gives us an over-determined view of social relations. Whilst a more productive reading has been suggested, its acknowledgement permits me to employ some licence for the purpose of continuity in this unfolding history. Homans may or may not be correct, but what of 'bringing back in' the other half of the population: women?

FEMINISMS: SOCIAL THEORY AND SOCIAL PRACTICE IN QUESTION

The potential for Enlightenment thought to marginalize whole groups of people was noted at the beginning of Chapter 1. Indeed, within feminist thought there exists a consensus that the intellectual and political cultures which have evolved over time have been invented by and for men. For some, the entire social universe of women's lives, including language and meanings, are viewed as saturated with male power (Marshall and Witz 2004).

There are two important issues: first, power imbalances shape theoretical construction; second, a group's place within the social structure influences theoretical attention they are afforded. For example, men in general remain the principle holders of economic and political power and this is replicated across history and the present. Men make up a large majority of corporate executives, top professionals and holders of political representation and mandate. Globally, men held 93% of cabinet-level posts in 1996 and most top positions in international agencies (Longino and Powell, 2004). In western culture in particular, men continue to control most technology such as the Internet (consider the major corporate 'owners' of Microsoft and Netscape – the two owners are men with a combined personal wealth of £97 billion – an amount which outstrips the combined GNP of most non-western nation-states) and weaponry (Smart 2006). With only limited exceptions it is men who have historically staffed and controlled the agencies of military force and judicial systems such as armies, intelligence agencies, police, prison and court systems (Connell 2000).

This section aims to bring attention to the ways in which theorists, in general, draw upon their 'background assumptions' (Gouldner 1971) to reproduce dominant ideas of men and women and from there, particular images of social relations. Biography and autobiography become important topics for this focus. For instance, in her study of housework, Ann Oakley takes Marx to task for the contradictions between the way in which he lived his life and the generation of his theoretical and political ideas (Oakley 1974). By not reflecting in this manner, social theorists and scientists in general are found to produce partial accounts and explanations which then masquerade as generalizations (Haraway 1991; Harding 1991). In the social sciences this has resulted in a lack of consideration of a whole range of core social activities fundamental to an understanding of social relations: 'areas like domestic work, or reproduction, or intimate relationships' (Oakley in Mullan 1987: 194).

Given the above, one aim of feminist theory is to focus upon an explanation for the exclusion of women from the considerations of social theory and social life in general. In the twentieth century this scholarship was inspired by those such as Virginia Woolf (1882-1941) and Simone de Beauvoir (1908-86) and then in what is known as 'second-wave' feminism. Writers within this genre include: Eva Figes (1978) Shulamith Firestone (1988) Betty Frieden (1965) Germaine Greer (1970) Kate Millett (1969) and Juliet Mitchell (1966).

From an historical viewpoint, feminist ideas were born out of the daily struggles of women in their everyday lives. These are then mobilized for the purpose of critically investigating the prevailing social order. Take, for example, the history of North American feminism. In the nineteenth century the women's movement aligned itself with the slavery abolition movement (Hole and Levine 1979). Sojourner Truth, once a slave herself, then illustrated the 'double jeopardy' that Black women faced when she spoke at a Women's Rights Convention in 1851:

> I think that between the Negroes of the South and the women of the North all talking about rights, the white men will be in a fix pretty soon. . . . That man over there says that women need to be helped into carriages and lifted over ditches, and to have the best place every-where. Nobody ever helped me into carriages, or over mud puddles, or gives me any best place . . . and ain't I a woman? . . . I have borne thirteen children and seen most all sold off into slavery, and when I cried out with my mother's grief, none but Jesus heard -and ain't I a woman?
>
> (quoted in Almquist 1979: 433)

Feminism takes particular aim at the alleged separation of theoretical justifications and social practices for the purpose of contributing to political change. This separation is often maintained for the purpose of enabling a 'distance' to be achieved in the analysis of society and social relations. From this arises notions of objectivity. Yet upon examination, the basis of this strategy is seen to rest upon unexamined assumptions concerning differences between the sexes. For instance, consider the culture-nature divide. In terms of the Enlightenment quest to differentiate ourselves from nature, women have been reflected in social and political thought via the view that culture is symptomatic of the 'rational man' and nature is closer to the 'emotional woman' (Marshall and Witz 2004).

Distinctions of this type are a topic for feminist theory. It is not suggested that biological differences do not exist between the sexes, but that these gain significance: 'within the framework of culturally defined value systems' (Ortner 1992: 489). Fused with their institutional reproduction within society, these ideas produce both social situations and images of women's place within society:

> the organisation of work, the division of labour, the ways in which political and social stability were envisaged, and the respective roles of husband and wife in childrearing . . . The divisions between male and female, science and nature, did not remain at the level of theorising but set the terms for a whole range of concrete struggles which are still being enacted today in the Western World.
>
> (Brown and Jordanova 1981: 239)

In these terms, the masculine becomes the sign of reason and is contrasted with the feminine as the sign of emotion. One is then a standard against which the 'other' is measured. As Lois McNay argues:

> A masculinist cultural hegemony is maintained through a process which involves the constantly practised differentiation of itself from what it believes it is not -the feminine.
>
> (McNay 1994: 37)

After all, we have found that a central way in which humanity has been considered is through the ability to exercise reason. In subtle ways, however, the 'knowers' of the Cartesian subject have been assumed to be men:

> Thus it would seem, if this assumption were right, that being emotional, or close to one's body and feelings, would mean being less rational, and therefore less free, and therefore less able to fully participate in all aspects of human life as a mature adult -less than fully human.
>
> (Griffiths 1988: 132)

In subjecting such ideas and practices to critical scrutiny, feminists seek to show that the ideas of 'men' and 'women' are social, political and historical constructs and not based upon so-called 'natural' differences.

According to Smith (1987) the cultural language and analysis of classical sociologists is that of men, male activities and experiences. For feminists, a major problem is that the classical 'scientific' definitions of the social world exclude large parts of human action and interaction (Marshall and Witz 2004). It presupposes a belief in individual difference and personal agency. It builds upon the concept of individuality developed in early-modern Europe with colonialism and capitalism (Gilroy 2001). This idea that men and women are qualitatively different did not exist until the 18th century and bourgeois ideology of separate spheres (Marshall and Witz, 2004). With the rise of 'Enlightenment', we witnessed images of 'white masculinity'. It was at this point that notions of reason, science, progress, and masculinity were merged into a unified concept of 'manhood'. Reason and objectivity also provided the moral legitimacy for the rise of capitalism and the modern organisation of society. For philosophers such as Kant (1773) reason tempered by science could overcome feelings and intuition. At the time of Enlightenment, utilitarian doctrines were gaining momentum enshrined by 'success is happiness': As industrial capitalism and the world of machines grew and flourished, this rationality included competition, planning, and goal attainment for men (Marshall and Witz 2004).

One aspect of the long history of modernity was the development of a separation between the public and private spheres. These had not always been separated in traditional societies, although there was usually a sex-based division of labour, often associated with a patriarchal system of male dominance. With the development of capitalism, cities, and industry, a public sphere dominated by men and male activities developed and expanded.

Women generally became restricted to the private sphere of household and family, and had limited involvement in political, economic, or even public social life. While some women were involved in more public activities, in the nineteenth century there were movements to restrict the participation of women in public life – for example, factory legislation and the family wage. Why? For Barbara Marshall and Anne Witz (2004) 'the family' remains an historical foundation for social theorists and at same time, for politicians, to legitimise inequalities and roles in modernity.

The notion of 'family' exists in two inter-related forms: first, 'familialism' – a widespread ideology of gender roles which run along lines of ideas of 'domesticity' and 'breadwinning'; second, as economic groups, the 'family' organises domestic and personal life providing functions to 'help' capitalism, for example, by providing child care, family care and domestic labour. The State has and still draws on not only capitalism to help it in its answers for production but also ideologies deriving from science in order to legitimise male and female roles and responsibilities to preserve existing social and moral order.

Women in both 18th and 19th centuries in England and North America were not recognised as individuals in legal structures or social theories (Smith 1987). Historically, Mary Wollstonecraft has been called the 'first feminist' or 'mother of feminism'. In her 1791 *Vindication of Rights of Man*, now considered a classic of feminist history, Wollstonecraft accepts the definition of her time that women's sphere is the home, but she does not isolate the home from public life. For Wollstonecraft, the public life and domestic life are not separate, but connected. The home is important to Wollstonecraft because it forms a foundation for social life and public life. For her, the State and public life enhances and serves both individuals and the family; men have duties in the family and women have duties to the state.

This also required an overhaul of concept of reason as a male dominated activity. In doing so, Wollstonecraft was also providing a critique of Enlightenment philosophers such as Kant who did not believe that such individual liberty was for women. Woman, for Kant, was incapable of reason, and only man could be trusted to exercise thought and reason. Thus, women could not be citizens. Yet Wollstonecraft makes her position clear: only when woman and man are equally free, and woman and man are equally dutiful in exercise of their responsibilities to family and state, can there be true freedom (Marshall and Witz 2004).

Coupled with this, in the 19th Century, Mill and Taylor, along with some early US feminists such as Elizabeth Cady Stanton and Susan Anthony, argued that the equality of women required full citizenship for women. This included giving women enfranchisement in 'democratic process'. After 1865, when Mill was in the English Parliament, he fought for women's suffrage. He also fought 'to amend the laws that gave husbands control over their wives' money and property'. He also supported the campaign for birth control information to be available, and was active in other campaigns that were aimed at assisting women and children (Smith 1987). Famously, the suffragette movement in early 20th century also added a key voice in pointing to the activities of women and striving for political action that represented women in political process and labour force.

The operation of patriarchy, defined broadly as the exercise of male power which serves to oppress women, has been reduced to the dynamics of capitalist relations. Yet feminist scholarship has shown that capitalism has particular effects not only upon women as a whole, but upon working class women in particular (Rowbotham 1973). This stands in contrast to the idea that the working class form a homogeneous group in relation to the means of production. From a historical viewpoint, it has been argued that the sexual division of labour actually predates capitalism (Barrett 1988**)**. Findings of this type thus throw doubt upon the ability of Marxism to explain the roles and positions of women in society.

Those feminists who seek to use Marxism for the purpose of such explanation have reacted to this state of affairs in differing ways. Some, whilst finding analytic mileage in writings, find it difficult to 'stick to the letter' of his works because the oppression of women, as a distinct class, did not concern him (Delphy and Leonard 1992). Overall, the necessity of modification renders the marriage between Marxism and feminism somewhat 'unhappy' (Weinbaum 1978). Either it should become 'healthier' or it will result in 'divorce':

> The struggle against capital and patriarchy cannot be successful if the study and practice of the issues of feminism is abandoned. A struggle aimed only at capitalist relations of oppression will fail, since their underlying supports in patriarchal relations of oppression will be overlooked.
>
> (Hartmann in Humm 1992: 109)

Whatever the ways in which feminist theories manifest themselves, it is clear that in meeting mainstream theory there is general agreement over the necessity for critique, modification or abandonment, depending upon the perspective from which the relationship is viewed. Therefore, in the encounter between Marxism and feminism new ideas have emerged. Take, for example, socialist feminisms which comprise:

> a set of interventions by feminists into socialist and especially Marxist theory, which aims to analyse and end the oppression of women in capitalist society. As such Socialist Feminism is built upon key texts, writers, political events, and organizations which comprise the Marxist tradition. It is these which have helped shape the agenda for Socialist Feminist debates and reformulations of Marxism.
>
> (L. C. Johnson 1990: 304)

Similarly, out of a confrontation with Freud's work, psychoanalytic feminisms have developed whilst within the writings of radical feminists two themes, in particular, become apparent. First, women as a distinct social group are oppressed by men, where patriarchy is taken to be *the* structure of oppression. Second, given the problems of incorporating male theories for the purposes of developing feminist ideas, radical feminism is seen to be created by women for the purposes of their liberation: 'So reproduction, marriage, compulsory heterosexuality, and motherhood are primary sites of attack and envisaged positive change' (Rowland and Klein 1990: 273). Given this, differences between Marxist, Socialist, radical and psychoanalytic theories frequently centre upon whether ideas generated by men can be used to explain the position of women in society. This, in turn, clearly relates to the political question of what is required in order to achieve women's equality. The theoretical and political are thereby fused.

Making Connections with Part II

The development of feminisms has necessitated a recovery of women's ideas from their historical marginality (Spender 1983). Further, there exists a continual need to maintain a link between theory and practice for the purposes of social, political and economic change. In its historical unfolding over several hundred years, feminist ideas have worked towards this end in a variety of ways.

In the first instance, there was the call for an incorporation of women into the changes taking place with the rise of modernity. In the works of both Wollstonecraft and de Beauvoir, for example, reason and liberalism combined in order that women may gain equality with men. However, on whose terms? If this is to utilize existing legislation, then are laws to be seen as neutral instruments? This is a highly problematic assumption given the nature of patriarchal relations. Here we witness another direction being formulated within the broad feminist agenda. A stress upon a more gendered view of social relations also appears which:

> glorified differences between men and women . . . Emphasis was placed on what is essential to women; by seeing the world through a woman's eyes, by allowing for a women's standpoint, we see not only a new perspective but one that is significantly better, more inclusive, and more life affirming. The celebration of women's traits, when linked to the idea that the marginalized and oppressed have an interpretive edge, gives women an epistemological or epistemic privilege.
>
> (Farganis 1994: 23)

The aim of this new generation of ideas was to erode 'the old distinctions between men and women' (Mitchell 1986: 45). This, as discussed, took place via a critique of the very grounds upon which dominant modes of thought and practice had based themselves. For some, this necessitated concentrating on patriarchy as the oppressive mechanism to the exclusion, for example, of the economic sphere. For others, as will become apparent in chapter 9, this emphasis was too one-dimensional.

What we now find is a terrain of feminist social theory which has been influenced by a rich history of ideas. In some cases, paralleling a current trend in social theory, the recognition of differences and diversities between women has led to the abandonment of any explanatory 'meta-narratives' which seek to account for the experiences of all women. The essential category 'women' as a social class and their relationship with men then gives way to an examination of the differences between women. Once the coherence of women's identities are questioned in this way, it has ramifications for any political movement which assumes a degree of homogenous characteristics amongst a group which may act as a unifying catalyst for change. Many of the contributors to one feminist reader thus question the use of Enlightenment rationality for the development of feminist theories in general (Marshall and Witz 2004).

To the contemporary theoretical landscape of socialist, radical, Marxist, liberal and psychoanalytic feminisms, we must now add both postmodern and poststructuralist feminisms. Given this, it is accurate to talk of isms', not feminism (Humm 1992). Therefore, in chapter 9 we shall outline the development and modification of feminist ideas since the mid to late 1960s.

This history will then be continued in chapters 10 and 11 where the issues raised by poststructuralist and postmodernist social theories are examined.

SUMMARY: THE LEGACY FOR CONTEMPORARY SOCIAL THEORY

In the beginning of this journey we found Dilthey's hermeneutics as a neo-Kantian defence of the dignity of the human sciences. Dilthey, combined with Hegel placed history firmly within the realm of social-theoretical development. This was a 'turn' in the Enlightenment project because it focused social theory on human understanding and away from a more positivist orientation on the externally observed regularities of human behaviour.

Phenomenology then entered our intellectual terrain and found us travelling from Husserl's transcendentalism to Heidegger's existentialism. Husserl, along with Dilthey, worked on an Enlightenment-based epistemological problematic. The result was that he prioritized 'consciousness' and added to the more subjectivist and action-oriented approach to social theory. This views society as:

> the creation of its members; the product of their construction of meaning, and of the action and relationships through which they attempt to impose that meaning on their historical situations.
>
> (Dawe 1970: 214)

Heidegger then subjected the divorce of knowledge (epistemology) from 'being' (ontology) to critique. In following the implications of this position social theory could no longer assume a unitary 'subject' who stands at the centre of the social universe. This is manifested in social theory in a number of ways. The combination of Heidegger and Nietzsche can be argued to have provided for the postmodern turn in social theory (Habermas 1992a) whilst Heidegger provides for a more ontologically focused social theory. This also affects how the 'subject' is conceptualized which, up to this point in time, had largely been assumed as something 'fixed':

> From Descartes's cogito, to Kant's and Husserl's transcendental ego, to the Enlightenment concept of reason, identity is conceived as something essential, substantial, unitary, fixed, and fundamentally unchanging. Yet other modern theorists of identity postulate a non substantiality of the self

> (Hume), or conceive of the self and identity as an existential project, as the creation of the authentic individual (Nietzsche, Heidegger, Sartre).
>
> (Kellner 1993: 142)

This challenge to the subject, combined with cultural and economic changes and critiques from other traditions of thought, has led social 'identity' to become a core topic in modern social theory (Calhoun 2000). Heidegger's critique of subject-centred consciousness also questions the mind-body duality which, since Descartes's time, had often been taken-for-granted. Together with the insights of feminist theory on the divide, this has added to an interest in the 'body' within social theory (Powell and Wahidin 2006). Nevertheless, this does not imply the acceptance of biology as a focal point for its development.

Moving away from the German-idealism that informed Husserl and Dewey's writings, the pragmatists tackled the subject-object dualism through concentrating on the inextricable link between consciousness and action in everyday life. Therefore, although developed in the United States, the potential existed for an alliance with both Nietzsche and Heidegger in a critique of the Enlightenment. This inspiration, from variants of pragmatism in our age of uncertainty, have moved the focus away from a search for foundations for social theory and ideas of 'absolute truth', to the notion that truth is dependent on what is 'useful' for action.

In another direction, phenomenology was found to stress a humanism through viewing the human subject as the source of knowledge (humanism also denotes secular atheism. Marx stresses a dialectic between human agency and material circumstances. At one level this gave rise to a humanist-Marxism, whilst the issue of structural or system determination versus a more action-oriented social theory, informed the work of the critical theorists who sought to recover the Hegelian and emancipatory aspects of his work. In so doing, they had to encounter thinkers who had challenged Enlightenment thought. In seeking to illuminate a number of central problematics, they needed to overcome the materialist-idealist dichotomy in social thought and take account of both objective and subjective factors in social life. This was achieved without a resort to consciousness by utilizing Freud's work. This also represented a critique of Cartesianism. In the resulting synthesis of ideas, an increased scepticism over the potential for emancipation within the Enlightenment project became evident via the argument that reason and rationalization were at increasing odds with each other.

Marx has also been read as a structuralist. As a general tradition of thought this travelled in two directions. First, there were the works of Saussure and Levi-Strauss both of whom worked outside of the Marxist tradition. Their ideas had implications for the study of social life through, in particular, a dissolving of the 'subject'. At the same time a neo-Kantian theme could be traced in the ideas of Levi-Strauss. Second, there was Althusser who, in returning to work, argued that there was an 'epistemological break' in his writings which provided for a 'scientific' Marxism.

A neo-Kantianism was also found in the work of Parsons. He perceived a convergence in social thought at the turn of the century and sought to overcome a dichotomy between positivism and utilitarianism in order to develop an action-oriented social theory. Whilst the subject of sustained criticism, Parsons brought the Anglo-American and continental traditions of social thought together. His ideas were then traced in the work of Luhmann, Habermas, Merton and Alexander. More generally, they have acted as an intellectual grinding wheel against which other social theories have sharp-ened their insights.

Feminism then entered this journey of ideas. Here was a body of thought whose beginnings reach far back into history. Yet these are the very voices that have been eclipsed through a series of unexamined presuppositions, replicated in everyday discourse, which are assumed to exist between the sexes. The implications of feminist critiques are wide-ranging. In particular, scientific thought and practice both reflect and justify a view of society which often excludes from consideration half of the population. This also opens up questions concerning the whole relationship between social theory and masculinity (Connell 2000) as well as the ways in which women and non-white peoples are excluded from dominant modes of thought and practice.

Collectively, these are a heady brew of ideas with which to travel into the terrain of contemporary social thought. Nevertheless, some clear themes have emerged. There is the question of the focus of social theory. Should it explain society without recourse to understanding the subjective interpretations of individuals, or should it seek to understand these everyday meanings, or can it achieve both?

With this general theme in mind, our aim is now to examine the terrain of contemporary social thought.

Chapter 4

FOUCAULT

What analytic tools does Foucault use for social theory? Once we have examined these, we can then consider his arguments on power and the subject. Unless otherwise stated, all references in the following two sections are to works by Foucault.

BEYOND STRUCTURALISM AND HERMENEUTICS

In order to situate Foucault's thinking, we need to understand the background against which he formulated his views. If we return to an earlier discussion on the intellectual climate in post-1945 France, two views emerge in relation to Husserlian phenomenology. First, structuralism, and second, a hermeneutics which did not resort to the transcendental subject, but argued that the source of meaning production should be sought in the social practices and texts which are fundamental to human activity. For the purposes of this section, we may trace this latter intellectual lineage from Heidegger. In the process, we shall be required to consider the generation of meaning without reference to an autonomous subject and to utilize the idea of discourses without allusion to underlying rules of structure. In other words, we have to move beyond the limitations of structuralism and hermeneutics in order to develop an 'interpretive analytics' (Dreyfus and Rabinow 1982; May and Powell 2007).

The fusion of these different ideas in Foucault's work results in an approach to the history of thought that is freed: 'from its subjection to transcendence' (1989: 203). In this journey we are also required to take two further steps. First, the relationship between the subject and truth should be viewed as an effect of knowledge itself. As Foucault put it:

> what if understanding the relation of the subject to the truth, were just an effect of knowledge? What if understanding were a complex, multiple, non-individual formation, not 'subjected to the subject', which produced effects of truth?
>
> (Foucault in Elders 1974: 149)

Knowledge should not, therefore, be viewed as 'theoretical' separate from the realm of 'practice'. Instead, knowledge becomes a practice which has the effect of constituting particular objects -non-theoretical elements - which are part of practice itself. Similarly, the theory of knowledge and the subject of knowledge become fused and the subject and truth are not related in the ways suggested earlier, but are seen as part of a relationship between knowledge and power that is socially constructed (Gutting 2003):

> The important thing here, I believe, is that truth isn't outside power, or lacking in power: contrary to a myth whose history and functions would repay further study, truth isn't the reward of free spirits, the child of protracted solitude, nor the privilege of those who have succeeded in liberating themselves. Truth is a thing of this world: it is produced only by virtue of multiple forms of constraint. And it induces regular effects of power. Each society has its of truth, its 'general politics' of truth.
>
> (Foucault 1980: 131)

What Foucault is doing here is rejecting certainty in social and political life and arguing that there is no universal understanding beyond history and society. In the process a 'history of the present' is written, according to which history is not excavated to reveal a deeper meaning but, instead, is viewed as a 'battleground' between relations of power. Now this begins, albeit in a different guise, to look like structuralism: that is, to view events as manifestations of a deeper, underlying reality. How is this issue tackled in his work?

In commentating on those who have labelled him a structuralist, Foucault writes: 'I have been unable to get it into their tiny minds that I have used none of the methods, concepts, or key terms that characterize structural analysis' (1989: xiv). Fine! Beyond assertion, however, what are his justifications for this position? After all, the opening for the type of analysis we have considered so far occurred in Althusserian structuralism. Ideologies are seen as ways of constructing subjects and their perspectives on the world, all of which takes place against the backdrop of a Marxist model of social reality.

The first point we might observe here is that Foucault eschews grand theoretical projects of the Althusserian type. Perhaps the most simple way to view his relationship with structuralism is by examining his analysis of historical 'events'. What we find here is a refusal to read historical 'events' as manifestations of deeper social structures, or as located on particular levels, accompanied by a focus upon the seemingly marginal as indicative of relations of power. Events are thereby seen to differ in their capacity to produce effects:

> The problem is at once to distinguish among events, to differentiate the networks and levels to which they belong, and to reconstitute the lines along which they are connected and engender one another. From this follows a refusal of analyses couched in terms of the symbolic field or the domain of signifying structures, and a recourse to analyses in terms of the genealogy of relations of force, strategic development, and tactics. Here I believe one's point of reference should not be to the great model of language (langue)and signs, but to that of war and battle.
>
> (Foucault 1980: 114)

If Foucault does not reach for structuralism, neither does he resort to grounds in the universal preconditions of human understanding (Heidegger). What he calls genealogy, following Nietzsche, is an analysis of cultural practices without a reference point outside of history itself. The motivating question underlying this approach is: 'What are we today?' (1988: 145). For it is history which makes us what we are and what we think of ourselves as being. This is a very different starting-point from, say, an analysis of the limits which are continually placed upon what is seen as our 'hidden' creativity as found in a critical social theory based in philosophical anthropology (Honneth and Joas 1988). The manifestations of such a project we have covered in relation to a 'dialectic of freedom' whose aim is to determine the good and bad elements in the Enlightenment. Genealogy, in contrast, does not involve a search for any 'origin':

> On the contrary, it will cultivate the details and accidents that accompany every beginning; it will be scrupulously attentive to their petty

> malice; it will await their emergence, once unmasked, as the face of the other. The genealogist needs history to dispel the chimeras of the origin.
>
> (Foucault 1984: 80)

We have come across other universal grounds for analysis through the employment of modified forms of reason; none more powerfully argued than by Jurgen Habermas. Yet the issues motivated by such an approach are still couched in terms of the ability to provide grounds for distinguishing between the true and the false, not in terms of the formation of truth itself. It is via this focus in Foucault's work that we uncover the legacy of Nietzsche.

Unlike Kant, Nietzsche did not ask what the grounds or basis for our knowledge was, but why knowledge was necessary in the first place. He argued that it was the belief we have in knowledge that is of importance, not whether it is true or false. This belief is seen to be indicative of a 'will to truth' which, as part of the 'will to power', involves the desire to affirm life as it appears to us. Truth now becomes inverted to be a fiction that is 'invented':

> What urges you on and arouses your ardour, you wisest of men, do you call it 'will to truth'? Will to the conceivability of all being: that is what I call your will! You first want to make all being conceivable: for, with a healthy mistrust, you doubt whether it is in fact conceivable. But it must bend and accommodate itself to you! Thus will your will have it. It must become smooth and subject to the mind as the mind's mirror and reflection. This is your entire will, you wisest men; it is a will to power; and that is so even when you talk of good and evil and of the assessment of values.
>
> (Nietzsche in Hollingdale 1977: 224)

With the Nietzschean turn in place, the object of Foucault's analyses became the production of discourses that provide justifications for actions: 'my problem is to see how men govern (themselves and others) by the production of truth' (1991c: 79). With a focus upon how the truth is constructed in a manner that permits, from the point of view of practice, a distinction to be made between the true and the false, Foucault's use of language as an analytic resource becomes distinct from any we have come across before.

Thus far we have considered the role of language in social life in a number of ways: De Saussure's structuralism; as a 'form of life' (Wittgenstein) analysed in terms of speech-acts (Austin, Searle); a fusion of speech-act theory onto the problematic of rationalization (Habermas) and the relationship between utterance and institutional authority (Bourdieu). We could undertake genealogical analysis by, for example, examining the rules of language governing an utterance and then derive a set of rules which produced the utterance in the first place. Our results may then be transposed from context to context. Once again, however, Foucault throws a spanner into the works of these approaches.

Foucault's analysis of discourse does not contain the idea of knowledge production via a meaning-producing subject, or rules that govern speech-acts, nor an allusion to rationality that poses some absolute value in reason (1991c). Given this, his approach cannot be situated within any of the above positions. Nevertheless, he sees discourses as a unity of statements in what is termed a 'discursive formation'. The question is: what provides their unity? Foucault rejects four possible answers to this question. First, that there is a well-defined field of objects to which they refer; second, that they possess a clear normative basis; third, that they possess what he calls an 'alphabet of notions'; and finally, that they are characterized by 'the permanence of a thematic' (1989: 37).

With these possibilities rejected, what we uncover is the existence of possibilities and differences in structures and rules. Where it does become possible to describe these 'systems of dispersion' in terms of objects, statements, concepts or themes that make up a 'regularity' then we may say:

> for the sake of convenience, that we are dealing with a discursive formation ... The conditions to which the elements of this division (objects, mode of statement, concepts, thematic choices) are subjected we shall call the rules of formation. The rules of formation are conditions of existence (but also of coexistence, maintenance, modification and disappearance) in a given discursive formation.
>
> (Foucault 1989: 38)

In a description of the events of discourse the question then becomes 'How is it that one particular statement appeared rather than another?' (1989: 27). However, there exists the potential for terminological confusion in Foucault's idea of 'statements'. Dreyfus and Rabinow thus propose the term 'serious speech acts' (1982: 48). This neatly encapsulates Foucault's adherence to the study of relations between speech acts and their continual formation and transformation where they are validated by particular procedures and the increasing army of 'experts' which characterize modern societies. It also takes account of Foucault's acceptance of the idea we have come across before to the effect that speech-acts are formulated against a background of everyday assumptions, whilst the idea of 'serious' retains his emphasis upon the generation of meaning within discourses.

With these themes and arguments in mind, we can now say that Foucault did not resort to a depth-hermeneutics whereby, via an excavation of the background assumptions which inform everyday interpretations, a 'truth' is revealed to exist. Further, that he was not content to view meaning as ultimately residing in a transcendental subject, nor to eliminate meaning through the employment of structuralist insights (Layder 2006). His concerns were to empirically examine the effects of social practices without recourse to some idea of their intrinsic homogeneity. It is to an understanding of this project in relation to power and the subject that we now turn.

Power and the Modern Subject

The above situates Foucault's approach to analysing social phenomena in terms of the politics of truth, but it says little of the effects of these discursive regimes. For this purpose, Foucault turned towards analyses of the exercise of power in contrast to a consideration of the forms of knowledge which create a sense of social order. In more specific terms, his objective became 'to create a history of the different modes by which, in our culture, human beings are made subjects' (1982: 208).

To return to our earlier discussion. The position which the analyst occupies in this process of inquiry is still not that of the 'bearer of universal values' (1980: 132). It is, however, as someone who is situated within society and does not speak for the truth, but engages in a 'battle' regarding its status and the political and economic role that it plays in social relations (1980: 132). In other words, what is taken as being self-evident in the present, is viewed as part of an historical process. From this point of view Foucault's (1991d) study on imprisonment and punishment aimed:

> to write a history not of the prison as an institution, but of the practice of imprisonment: to show its origin or, more exactly, to show how this way of doing things - ancient enough in itself - was capable of being accepted at a certain moment as a principal component of the penal system, thus coming to seem an altogether natural self-evident and indispensable part of it.
>
> (Foucault 1991c: 75)

Three lines of inquiry are required in order to understand how the subject is objectified. First, how particular sciences contribute to this process. Second, a study of the 'practices' that divide people from others, as well as within themselves: for example, the mad and sane, the sick and healthy and the criminal and non-criminal. Third, in what ways people then turn themselves into subjects (Foucault 1982: 208). Just to reiterate: the form of this analysis is opposed to the search for origins and ideal significations, as well as the notion that we possess an intrinsically rational and coherent sense of self.

At this point we find another central aspect to Foucault's work which causes interpretative confusion. Quite simply, it is the tension which exists between his approach to the study of social relations and a deep-seated idea within Western cultures: that is, the belief that the condition of freedom is the absence of power and thus power is always repressive. To commit oneself to this tradition would be to fall into the trap that Foucault's inspired genealogies seek to avoid. For instance, when it comes to the individual in relation to the means of production, or as involved in relations of signification, we turn to economic history, linguistics and semiotics to shed light on these matters. When it comes to power all we ever ask is: 'What legitimates power?' (1982: 209). However, with the individual implicated in these relations, this approach cannot suffice. In order to understand how the subject is objectified, this idea of power needs total revision. This is where Foucault's 're-conceptualization' of power is of central importance and we need to understand what he means by this before we continue with our discussion.

Power is not inevitably vested in the state, nor localized in, say, relations between the citizen and state. Instead, the effects of the domination of power are attributable to various 'dispositions, manoeuvres, tactics, techniques, functionings' (1991d: 26). Its properties are invested in those over whom it is exercised, so it not only exerts pressures upon people, but is also transmitted by them: 'Power comes from below; that is, there is no binary and all-encompassing opposition between rulers and ruled' (1979: 94). Power is part of the social body as a whole -in the machinery of production, families, limited groups, and institutions (1979: 94). Thus, power is not a property of capitalism, patriarchy, or the bourgeoisie - it is a strategy that has evolved as part of the characteristics of modern society. Nor should power be identified with an individual who possesses or exercises it by right of birth:

> Not the domination of the King in his central position, therefore, but that of his subjects in their mutual relations: not the uniform edifice of

> sovereignty, but the multiple forms of subjugation that have a place and function within the social organism.
>
> (Foucault in Kelly 1994: 34)

Given this, if we conceive of power as emanating from one source, we will not understand its effects. When it comes to politics and power, the sovereign's head must be cut off. Further, to think of power as repressive (the repressive hypothesis) is to neglect its positive aspects. In this sense, power becomes 'a machine in which everyone is caught' (Foucault 1980: 156).

Yet a question still remains: if power is transmitted in this way, where does it come from in the first place? This is one of the most difficult concepts to grasp. After all, if knowledge and power are linked, one thinks of Marx and the idea that those who are in charge of the means of production are, at the same time, in charge of the means of mental production. An illuminating discussion of this issue can be found in an interview with Foucault, conducted by Jean-Pierre Barou and Michelle Perrot (Foucault 1980: 156)

Let us take the example of a factory: here, relations of power may be seen in terms of how 'individuals try to conduct, to determine the behavior of others' (1991b: 18) The system of power is pyramidal and it occurs between the managers and the managed. However, the strategic apex (the managers) is not the source of power, nor is a principle or goal which organizational members are expected to invest with legitimacy in terms of their practices. Power and its techniques are used throughout the factory. Therefore, it cannot be simply identified with a particular group:

> These tactics were invented and organised from the starting points of local conditions and particular needs. They took shape in piecemeal fashion, prior to any class strategy designed to weld them into vast, coherent ensembles. It should also be noted that these ensembles don't consist in a homogenisation, but rather a complex play of supports in mutual engagement, different mechanisms of power which retain all their specific character.
>
> (Foucault 1980: 159)

We can draw three points from this quote. First, in his analysis of power Foucault is concentrating on the local and contingent. Therefore, as noted before, he is not prepared to 'read off' practices against some universal idea of rationality. Rather, he is concerned to examine the ways in which particular rationalities are deployed, leaving the connections to be established amongst them open to investigation (Dean 2007). Second, power should not be seen, from this point of view, as existing above, but alongside developments in the forces and relations of production. Power is transformed along with, but not by these changes and may be found in both socialist and capitalist societies (1980: 160). Third, the workings of this form of power can be traced far back into history.

In terms of tracing the history of this form of power, it was during the seventeenth, eighteenth and nineteenth centuries that a new form of political rationality found its target. When faced with the issue of population growth, political economists examined it in terms of wealth accumulation and the productive capabilities of labour, with the causes of poverty being conceived of in terms of disease, idleness, etc. The social reformer and philosopher Jeremy Bentham (1748-1832) on the other hand, 'poses the question in terms of power - population as object of relations of domination' (1980: 151). It was the economic changes occurring in the eighteenth century that:

> made it necessary to ensure the circulation of effects of power through progressively finer channels, gaining access to individuals themselves, to their bodies, their gestures and all their daily actions. By such means power, even when faced with ruling a multiplicity of men, could be as efficacious as if it were being exercised over a single one.
>
> (Foucault 1980: 151-152)

The differentiation of the population thus required this new form of power. The simple domination of one group by another was no longer a feasible technique given the increasing complexity of society. Contrary to simple ideas of coercion, power has both a political and economic cost. If you are too violent in controlling a population, you risk revolt. If you intervene too frequently, then resistance and disobedience may result. Disciplinary technologies and regulatory procedures must inform and produce a normative basis which spreads throughout the social body, whilst masking the operation of power that underlies this process (Katz 2007). In this way, the modern subject becomes inextricably linked to the society of which they are a part through turning themselves into an object:

> There is no need for arms, physical violence, material constraints. Just a gaze. An inspecting gaze, a gaze which each individual under its weight will end up interiorising to the point that he is his own overseer, each individual

> thus exercising this surveillance over, and against, himself. A superb formula: power exercised continuously and for what turns out to be a minimal cost.
>
> (Foucault 1980: 155)

The body, space and time are routinely controlled (Turner and Dumas 2006). What Foucault calls 'bio-power' relies upon the scientific categorization of the population, in terms of being an object of systematic and sustained political intervention, as well as encompassing a focus upon the human body as an object of control and manipulation. Collectively, these 'technologies' centre around the 'objectification' of the body. They form a 'disciplinary technology' whose aim is to forge the individual, with 'normalization' being one technique (Foucault 1977). This is best exemplified in enclosed situations, such as a prison, school, hospital or, as above, the factory, where the control of space and time is more readily apparent. It is here that we find:

> a whole micro penality of time (lateness, absences, interruptions of tasks), of activity (inattention, negligence, lack of zeal), of behaviour (impoliteness, disobedience), of speech (idle chatter, insolence), of the body ('incorrect' attitudes, irregular gestures, lack of cleanliness), of sexuality (impurity, indecency).
>
> (Foucault 1991d: 178)

This disciplinary technology permits supervision and control with interiori-zation being the most effective deployment of power. In this respect, Foucault suggests that there are three ways in which 'self-examination' has developed over time. First, the Cartesian form such that thoughts are considered in relation to their correspondence with reality. Second, the ways in which our thoughts correspond with, or relate to, rules. Third, the relationship between a thought which is 'hidden' and some inner impurity:

> At this moment begins the Christian hermeneutics of the self with its deciphering of inner thoughts. It implies that there is something hidden in ourselves and that we are always in a self-illusion which hides the secret.
>
> (Foucault 1988b: 46)

The discourse of constituting the subject as object thereby also requires the discourse of the subject and this too, has changed over the course of history. The means of exposing the truth of ourselves in Christianity involved being subjugated to a master. This model of obedience means that once the confession was verbalized, it was accompanied by a renunciation of the sense of self and 'will' as part of a continual process of self-renunciation: that is, disclosing oneself to an authority. This is a 'permanent verbalization'. However, what then occurs is that the interpretative sciences come to play this central role. After all, it is they who 'claim to be able to reveal the truth about our psyches, our culture, our society -truths that can only be understood by expert practitioners' (Dreyfus and Rabinow 1982: 180):

> From the eighteenth century to the present, the techniques of verbalization have been reinserted in a different context by the so-called human sciences in order to use them without renunciation of the self but to constitute, positively, a new self. To use these techniques without renouncing oneself constitutes a decisive break.
>
> (Foucault 1988b: 49)

This is accompanied by the state performing 'a modern matrix of individualization' (1982: 215). Together, these are the new forms of 'pastoral power'. Nevertheless, whilst claiming to speak the truth, these interpretative sciences are actually part of the rationalities which inform a political technology that saturates everyday life (Dean 2007). Importantly, however, this should not be conceived of as being a deterministic relationship.

To illuminate the indeterminacy of these practices and hence the possibilities for their transformation, the focus of Foucault's investigations changes. He now considers not so much the ways in which we are constituted as both subjects and objects, but the effects of this on our relations with others and the implications for the relationship between power, truth and the self. Now these are not discrete topics, but ultimately complementary strategies of historical investigation (Foucault 1988c: 15). The body then becomes not only the object or target of power, but also the centre for resistance and opposition. After all, contemporary society is increasingly characterized by struggles against the exercise of power:

> opposition to the power of men over women, of parents over children, of psychiatry over the mentally ill, of medicine over the population, of administration over the ways people live.
>
> (Foucault 1982: 211)

Foucault asks the empirical question 'what happens' when power is exercised? As opposed to becoming involved in endless debates over what power 'is' and where it comes from (1982: 217)? Resistance to forms of power is the starting-point for this investigation. These are struggles for new subjectivities which encompass the right to be different and distinguish oneself from the ways in which the effects of particular 'technologies of the self' routinely constitute people: 'These struggles are not exactly for or against the "individual", but rather they are struggles against the "government of individualization"' (1982: 212). This makes sense if we remember that the individual is not constituted before the effects of power because:

> it is already one of the prime effects of power that certain bodies, certain gestures, certain discourses, certain desires, come to be identified and constituted as individuals.
>
> (Foucault 1980: 98)

POSTMODERNISM

Postmodernism not only affects culture as a whole but a new social epoch emerged due to a range of developments in all areas of society such as politics, industry, media and the rise of social movements (Layder 2007). This has ultimately has led to the development of a postmodern world characterised by 'fragmentation, multiplicity, plurality and indeterminacy' (Thompson 1992: 223). If this characterisation is correct, it has had strong implications for the Enlightenment aims and values. Within social theory, Zygmunt Bauman claims that postmodernity is, above all, a state of mind (Bauman 1992: vii). He suggests that the term 'modernity' has come about only through the postmodern vantage point. The labelling of modernity is merely a re-evaluation of the past and is not a term which is simply descriptive of a certain period in the logical progression of history (Lemert 2006). This is equally true of postmodernity in that it is not merely a label for a set period in time. Postmodernism represents a shift of focus from universalism to pluralism. It is the acceptance that modern aims of universalism are futile and the recognition of 'pluralism of cultures, communal traditions, ideologies, 'forms of life' or 'language games'' (Bauman 1992: 102).

Chapter 5

THE EMERGENCE AND ANALYSIS OF THE POSTMODERN

THE RISE OF THE POSTMODERN

Certainty and progress is assumed to characterize the epoch known as modernity. An emerging attitude became apparent. This may be called the 'attitude of the modernists'. These were and are the people who celebrate the potentiality which is said to exist in history, art, science and politics and the achievements of the human race.

This attitude, therefore, is one of ambivalence in its viewing of broken promises and celebrations (Bauman 2001). We have found those who approached their analyses of social relations in this manner: for example, Simmel and Weber. Their writings can be contrasted with the relative optimism of Marx and Durkheim. Weber regarded Western societies as exhibiting an increased potential for control in the name of instrumental rationality. A fatalism was then apparent in his reflections via a recognition that the realm of human meaning was being trammelled by vast impersonal forces. Weber was not prepared to overcome such observations through an allusion to the external by the freezing of time and all of its 'fleeting qualities' (Harvey 1990: 21). From this point of view, we can read both Weber and Simmel as writers who were straddling a chasm that was opening up between two epochs.

A recognition of this state of affairs was apparent in the writings of Mills, who sits within the broad tradition of American pragmatism. He wrote of the changing ideas of 'reason' and 'freedom' and the collapse of liberalism and socialism as explanations of ourselves and the relations into which we enter (there are echoes here of Weber's brittle synthesis of Nietzsche and Marx). The social sciences are seen to have inherited terms which, although outdated, are still part of its practices. If generalized to contemporary situations, they 'become unwieldy, irrelevant, not convincing' (Mills 1970: 184). He thus argues that the Modern Age is being superseded by a 'Fourth Epoch'. He refers to this as the 'postmodern period':

> *The ending of one epoch and the beginning of another is, to be sure, a matter of definition. But definitions, like everything social, are historically specific. And now our basic definitions of society and self are being overtaken by new realities . . . I mean that when we try to orient ourselves - if we do try - we find that many of our old expectations and images are, after all, tied down historically.*
>
> (Mills 1970: 184)

In particular, it is the role of reason and the idea of the free individual which he argues are the prime legacies that twentieth-century social scientists have inherited from the Enlightenment (Mills 1970: 186). In the face of change, the central issues that now face us are concerned with our very nature and our limits and possibilities. After all, is our destiny to become 'what may be called The Cheerful Robot?' (Mills 1970: 189)

We may characterize this tension as one which exists between two elements of the process of modernization: specialization and rationalization. Specialization, as Durkheim noted, necessitates a complex division of labour which, for the purposes of social cohesion, requires a new form of social integration. This time, of course, it was with society taking the transcendent role, not the autonomous subject. Authority and freedom, therefore, were not opposites. For Weber, the same process was charted but he then wrote of 'specialists without spirit and sensualist without heart'. This was symptomatic of the process of increasing societal complexity running alongside greater centraliz-ation and control.

Habermas (1992) claims that the process of societal evolution depends upon a normative basis which exists in the lifeworld, but in the forward march of modernity the 'steering media of power and money' evolve which have no basis in the lifeworld. As we have seen, for there is a way out of this 'iron cage' through a modified Enlightenment project which finds the possibility for a consensus that is not imposed upon the lifeworld, but emerges from within it. However, is this quest for consensus yet another version of the modernists' attitude which is doomed to failure? Do meta-narratives of this type, in their desire to universalize, render violence to heterogeneity? This is clearly the case for Lyotard (1924-1998) who, in reference to Habermas, plays on the same themes that the earlier quote indicated:

> Is legitimacy to be found in consensus obtained through discussion, as Jurgen Habermas thinks? Such consensus does violence to the heterogeneity of language games. And invention is always born of dissension. Postmodern knowledge is not simply a tool of the authorities; it refines our sensitivity to differences and reinforces our ability to tolerate the incommensurable. Its principle is not the expert's homology, but the inventor's paralogy.
>
> (Lyotard 1984: xxv)

A recognition of heterogeneity is not the monopoly of postmodernist perspectives (Ritzer and Goodman 2001). That noted, what seems to have changed in socio-theoretical discourse is that those who are the bearers of the dream of universalism now have to defend themselves against the relativists (Mouzelis 2007). In considering these issues we have to ask if this is symptomatic of a changing set of social conditions such that there can no longer be any unifying principles of co-ordination or communication that are available to us?

In considering this question, we might first recall the arguments of Kroker and Cook that Augustine, whilst signalling the beginning of the modern age, also possessed a discourse on the will that commenced the 'arc of a dead power which will not come fully into light until the nineteenth century in Nietzsche's nightmarish vision of the "will to power"' (Kroker and Cook 1988: 72). In these terms, they can then go on to characterize the postmodern condition as:

> a coming home to the 'perfect nihilism' (Nietzsche) which has always been at work in Western consciousness and which only now, in the fully realized technological society, reveals itself in the fateful meeting of power and the sign.
>
> (Kroker and Cook 1988: 73)

Nihilism may be viewed as the celebration of the relativism of values such that there is no absolute good, or the refusal to create ultimate values, or an absolute subjectivity that accompanies a 'loss of the world' (Glucksmann 1992: 336). Nevertheless, from whatever vantage point it is viewed, if the postmodern condition is that which gives rise to its celebration, then attempts at delineating a new basis for reason are not moving against some trendy intellectual movement, but 'real' historical changes that are taking place in the organization of societies as a whole. If this is the case, then consensus would, as part of the modernists' dream of a better future, indeed become 'a horizon that is never reached' (Lyotard 1984: 61).

As noted, the forward march of modernity came to witness the increasingly complicated nature of the social world being compensated for by mechanisms of communication and control. These functions were taken over by systems which came to represent the need for specialization and co-ordination. These systems are argued to reduce the complexity of the social world. However, does their legitimacy then come to supersede that of the realm of human values? For instance, even matters of risk, as part of both the excitement and tragedy of the human condition, are now open to their calculations:

> In binary coding, modern society possesses a highly specific form for heightening, normalizing and contextualizing risky behaviour. By attributing observations to certain coded systems, we can recognize which risk network we are operating within -and which one we are not operating within. Whenever a matter is dealt with in the context of a binary code, the implication is that not only the positive value but also the countervalue could be assigned validity.
>
> (Luhmann 1993: 76)

These systems come to represent the site of reason and it is from these that Habermas seeks to recover its critical capacity in order to restore the power of rational argumentation. Despite these differences, both systemic and critical modernism represent a celebration of the potentiality of reason to achieve regularity and understanding. Yet there is an historical process that might be argued to counter these attempts.

One such process is that of 'de-differentiation' (Lash 1990). The implications of this process being that a differentiation between the aesthetic, cultural, social, economic and political realms, as a characteristic feature of modernity, is no longer a feasible option for analytic purposes. Instead, from a cultural vantage point, we now have a colonization of the moral-political and theoretical spheres by the aesthetic; a collapse of the social into the cultural, accompanied by a breakdown of the barriers between high and low culture and finally, in terms of 'cultural economy', the 'death' of the distinction that was once made between the author and their text and, for instance, the distinction that is made between the production of a play and its reception by its audience who now form part of its cultural production (Butler 2003; Ritzer and Ryan 2007).

If we extend this process, beyond the cultural sphere to which Lash and Urry confine it (Lash and Urry 1987; 1994) it follows that we can no longer divide the social world into autonomous units. In this sense, there are those who are willing to engage in this embrace via the argument that differentiation extends to all spheres of social life. Whereas the process of modernization, as series of socio-economic changes, bore witness to the specialization of sub-systems by function such that those systems with similar functions possess similar structures, we are now said to be witnessing 'an extension of this process in which social units are differentiating at the level of structure as well as of function' (Crook, Pakulski and Waters 1992: 33). In these terms:

> Patterns of economic production may be said to be hyperdifferentiating, providing a multiplicity of directions in the absence of a unifying structural principle. Such hyperdifferentiation of structure is, we argue, a generalized and widespread pattern.
>
> (Crook, Pakulski and Waters 1992: 33)

Durkheim wrote: 'A discipline may be called a "science" only if it has a definite field to explore. Science is concerned with things, realities' (in Giddens 1972b: 57). The basis upon which this was to proceed is the correspondence between the language of representation (theory) and reality. Yet the process of de-differentiation renders this strategy obsolete. There is no 'reality' to represent, only realities. There is no reason to base analysis upon, only reasons.

A tension between the forces of centralization and dispersion under postmodernity now manifests itself in various ways. First, the media of control change. Second, as noted above, the ability to control the object world is lessened. Third, there is a collapse of the boundaries between the different spheres of social life: for example, the public and private. In relation to the first aspect, specialization is seen to be superseded by hyperdifferentiation whereby control is no longer the province of specific spheres of life, nor specialized disciplines for their study (Layder 2006). This all results in the generation of a 'limitless range of media types' within society:

> for example, TV churches can generate money as well as value-commitment and influence, and businesses which professionalize their employees and offer them lifetime security can be a source of value-commitments as well as money. In other words there is no longer an oligopolistic set of sources for money and power and they thus lose their effectivity as sources of control.
>
> (Crook, Pakulski and Waters 1992: 34-35)

Similarly, the breakdown of the correspondence between structure and function, in turn, leads to a breakdown of the potential for prediction and control: 'States become ungovernable, economies unmanageable, and worlds anarchistic' (1992: 35). The result being that postmodernization is characterized by unpredictability and chaos, not homogeneity and control. However, not only is system de-differentiation a characteristic of this process, but also when it comes to an analysis of social action, this is seen to be:

> divorced from underlying material constraints (or rather these constraints disappear) and enters the voluntaristic realm of taste, choice and preference. As it does so the boundaries between determined social groups disappear. So class, gender and ethnicity decline in social significance and so also do some of their characteristic forms of expression, including class-based political action and the distinction between high and popular culture.
>
> (Crook, Pakulski and Waters 1992: 35)

With this breakdown of divisions in social space, the cultural realm of signs and symbols comes to penetrate the socio-structural realm of social relations to such an extent that individual and multiple realities come to predominate (Layder 2006). Therefore, the potential for individual distinctiveness is heightened. At this point in our journey, it is the ideas of Jean Baudrillard which become of central importance.

Postmodern Social Theory

Baudrillard (2004) argues that we have moved from societies which are characterized by production, to those of consumption. His ideas have been charted as moving from a development of Marxism, through a transitory phase, before finally becoming a critique of Marxism itself (Butler 2003). For instance, in relation to the object world, Baudrillard was to argue that the things we consume form systems of signs. These systems, to recall our earlier discussions on Saussure, are self-referential and precede the individual. Now, during the earlier stages of his work, whilst seeing consumption as determined by production in a classical Marxist manner, the influences of Barthes and Mauss also provided the basis for a structuralist analysis of symbolic exchange. Ultimately, this was to render distinction between use and exchange value untenable. Briefly, what form did this transition take?

In Marxist terms, a commodity is seen as having a basis in a reality through the meeting of some specific human need. Adding the insights of French structuralism to this, leads one to see commodities as objects of desire that then enable the individual to relate to the social order via their consumption. In this way, people may be said to be both ordered and differentiated. However, Baudrillard now starts the process of bringing that which is considered to be marginal in Marxist thought, to having analytic primacy. In order to understand this we need to note the distinction that Marx made between exchange and use value in relation to commodities (Ritzer and Ryan 2007).

Use value is said to relate to the utility that a commodity has in relation to satisfying a human want. Exchange value, on the other hand, possesses value due to its ability to be exchanged for other goods or services. Therefore:

> The use-value and the exchange-value of a commodity are not just two different determinations, or aspects; they are contradictory determinations. A commodity is a use-value only if it is immediately useful to whoever has it. It is an exchange-value only if it is not immediately useful but is used for exchange to get something else.
>
> (Cleaver 1979: 90)

Exchange-value is the opposite of use-value. From a political vantage point, use-value may be related to working-class commodity consumption which is utilized for the satisfaction of needs. Capitalism, on the other hand, is concerned with exchange values for the extraction of surplus value and profit. Despite this, each depends upon the other, for example: 'It is exactly because labor-power is a use-value for capital that it is an exchange-value for labor' (Cleaver 1979: 92). Nevertheless, for Marx, use value still functions as a basis for the objective assessment of needs that are assumed to exist independently of relations of capital (Lavalette and Pratt 2007). Yet how can one evaluate and compare in this manner separate from the social context of consumption?

In contrast to this approach Baudrillard argues that use-value should be viewed in terms of structures of symbolic exchange:

> consumption does not arise from an objective need of the consumer, a final intention of the subject towards the object; rather, there is social production, in a system of exchange, of a material of differences, a code of significations and invidious (statuaire) values.
>
> (Baudrillard 1993: 75)

This is not a system of consumption whereby the individual subject makes a rational choice, through calculation, in terms of what to consume. Instead, meaning is located within a structure of differences that are conferred upon objects. Take, for example, the consumption of leisure. Is this about the need of the autonomous subject to enjoy 'free time'? No! It is the 'consumption of unproductive time' (Baudrillard 1981: 76). What does this mean? Consider the man [sic] of leisure:

> Outside of the competitive sphere there are no autonomous needs. Spontaneous motivation doesn't exist. But for all that, he can't permit himself to do nothing. At a loss for something to do with his free time, he nevertheless urgently 'needs' to do nothing (or nothing useful), since this has distinctive social value.
>
> (Baudrillard 1981: 77)

Analytic primacy is afforded to the relationship between value and symbolic exchange, not use and exchange value. Use value has now disappeared within a system of sign values: 'This must be recognised today as the truth of the sphere of "consumption" and the cultural system in general' (Baudrillard 1981: 87). Indeed, science, artistic and intellectual work and even 'transgression' and 'innovation' are part of the same process (Gane 2006).

Baudrillard could choose to analyse the relationship between the workings of the system and the cultural processes that inform it - as had Marx before him (Gane 2006). However, having asserted primacy to the system of symbolic exchange, he then hovers between seeing the system from the point of view of an oppressed class or group and as a system of signs of consumption which refer only to each other and have no external referent. In arguing for the latter, production is to give way to consumption with its vortex of symbols. As such, any hint of a humanist legacy with respect to a state of 'alienation' is abandoned. At this stage in his thought, he thus exhibits a curious blend of Marx, Durkheim, Nietzsche and Freud in the development of what Mike Gane characterizes as a 'non-humanist idea of alienation' (1991: 95).

All of this was taking place in the era of the 1970s. It was then possible to believe in the theoretical and political possibilities that came with a fusion of Freud and Marx: 'it was in their copulation that things could be saved' (Baudrillard in Gane 1993: 59). This was not to last long with Nietzsche hovering in the background. Here was a thinker who Baudrillard acknowledges is the most influential figure in relation to his thought: 'but not as a point of reference . . . but as a spirit (esprit), as a stimulus (impulsion), or inspiration' (Baudrillard in Gane 1993: 203)

With Nietzsche in place, Baudrillard will now take us into the condition where models of reality, and not reality itself, predominate. No longer is reality to be measured against the sign, nor are they seen to exist in a dialectical relationship. There is now only the sign:

> Now the sign seems to me to posit what I have called the 'principle of hyper-reality'. That is, what we have now is the disappearance of the referent -and it is in relation to this disappearance of the referent that there is a sort of omnipresence of the sign.
>
> (Baudrillard in Gane 1993: 142)

The system now comprises the consumption of signs without an external referent. Critical theory, for example, cannot allude to a dialectic of the Enlightenment to discover the truth with an existing state of affairs. The negation of reality and reality itself are now one: 'In short, what we now have is a principle of non-reality based on "reality"-a principle of hyper-reality as I call it' (Baudrillard in Gane 1993: 143). This is the era of postmodernity: nothing real exists, but only 'codes' of the original and it is these that enable simulation to predominate over production.

This is Baudrillard's final rejection of Marxism. The real and its representation have become erased to such an extent that everything is either, or has the potential to become, simulacra: that is, we only have models of reality and it is no longer possible to represent things as they are. The ability to decide, for example, between ideas, aesthetics and political persuasions, collapses. This is the era in which information is predominant. Computers, cyberspace, the media and information technological processing systems in general add to a 'politics of seduction' which are part of the signs that are a routine feature of our everyday lives:

> life is conducted in a ceaseless circulation of signs about what is happening in the world (signs about news), about what sort of identity one wishes to project (signs about self), about one's standing (signs of status and esteem), about what purposes buildings serve (architectural signs), about aesthetic preferences (signs on walls, tables, side-boards), and so on.
>
> (Webster 1995: 177)

There is no transcendental position from which one may view this process and pronounce upon its 'truth'. The best we might achieve is to be a spectator in the process of seduction. The sign is no longer to refer to the 'real' as a semiological strategy might seek. The steering media of modernity is now the media in postmodernity. What Weber referred to as 'disenchantment' is the irreversible postmodern condition in which meaning has 'imploded'. There is nothing beyond appearance, for appearance is all we have: 'The screen, the image, the sign, the modern message, the media require profusion, proliferation; there must be plenty for everyone' (Baudrillard in Gane 1993: 148). This is the end to history that leaves rationalism in its wake:

> We tend to forget that our reality, including the tragic events of the past, has been swallowed up by the media. That means that it is too late to verify events and understand them historically, for what characterizes our era and our fin de siecle is precisely the disappearance of the instruments of this intelligibility. It was necessary to understand history when there was still history.
>
> (Baudrillard in Gane 1993: 160)

The political realm may no longer be viewed of as autonomous, whilst the social realm is now so saturated with media images that is has become 'anonymous'. It follows that those meaningful referents invoked during the course of social inquiries - class, gender, age, disability, ‘race’, ethnicity - disappear in the collapse of the 'social' (Ritzer and Ryan 2007). These concepts are no longer tenable because the real has been overrun by the 'hyperreal'. Allusions to characteristics of the 'masses' are fatally flawed:

> The mass is born short-circuited in 'total circularity'. What seems to occur he argues is a circulation around simulation models, and a collapse of the complex system into itself . . . For no-one, no organization as has been the case previously can any longer speak with confidence 'for' the mass.
>
> (Gane 1991: 136)

Baudrillard takes this into America. Here is the postmodern condition where culture and social life are not distinct. Europeans are said to locate themselves in terms of their past and their destiny. However, America is a 'fiction' where the notion of the 'authentic' which accompanies a sense of history disappears in a perpetual simulation in the 'present'. We find cities which are 'exponential', not 'referential'. Las Vegas is Disneyland. This is not a referential city: that is, one with a past, a defined territory and a sense of history. Instead, it is a city whose 'total superficiality . . . is bound up with superficiality of gambling and speculation itself' (Baudrillard 1993: 246). This is the imaginary of the real that is to be found in Disneyland:

> In both Disneyland and Disney World, it is clear that everything that can be derived from the imaginary has been caught, represented, made representable, put on display, made visual. Literally putting on a show for consumption without any metaphors is obviously a radical deterrent to the imaginary. Once again, utopia becomes reality.
>
> (Baudrillard 1993 : 246)

Baudrillard focuses then on the transcendental nature of society through the concept of hyperreality. He argued that in today's society there is no distinction between reality and illusion (Rojek 2004), and that individuals live their lives through a simulation of reality. Nothing has any true origin or authenticity and lived experience is itself a mere construction, made of a series of depthless signs and representations (Smart 1993: 52). In this way Baudrillard argues that not just consumer society but social sciences as disciplines of study can serve no political purpose anymore as power relations have been dispersed through the hyperreal nature of society (Smart 1993: 55). The 'neat divisions', 'hierarchies', and 'foundational premises' of both modernity and social science (Seidman 2004) are no longer relevant. In particular, Baudrillard (2005) rejects the economic determinism of Marxism (as is true of all the first wave postmodernists) and states that there are no longer such fundamental systems of exploitation, only superficial simulations and exchanges:

> Like the philosopher Nietzsche, Baudrillard criticizes such claims to truth and favours a model based on what he calls seduction. Seduction plays on the surface: it is the surface appearance that is effective in determining action, not some latent or hidden structure as claimed by Marxism or Freudianism.
>
> (Thompson 1992: 244.)

With the obliteration of the ability to allude to those processes, products, effects, forms of rational argumentation, rules, or relations of cause and effect which might enable us to explain and decide between truth and falsehood, we have entered the postmodern condition. This is the end of meta-narratives of explanation that accompanies 'an internal erosion of the legitimacy principle of knowledge' (Lyotard 1984: 39). This is yet another vantage point through which we can illuminate Lyotard's antagonism for any search for consensus within social life. Lyotard found it necessary to distinguish between what constituted pre-modern, modern and postmodern thought and he then set out to deconstruct the way in which bodies of knowledge are created in order to legitimate hierarchical structures in society.

In his influential work 'The Postmodern Condition' (1984, first published in French in 1979) Lyotard looked at the changes that have occurred to the nature of 'knowledge' throughout history. He pointed to how premodern society was based on narratives that were made up of religion and myth (Seidman 2004). Knowledge was a body of stories that were thought to explain the way society was and determine that which was 'good' or 'evil'. In this way such narratives legitimated the social rules of behaviour that determined how society was structured and who had authority (Seidman 2004). In contrast modernity, Lyotard argued, was thought to be based on 'true' knowledge that rejected the 'narratives' of premodernity. However in 'The Postmodern Condition', Lyotard asserts that in actual fact supposedly 'pure', 'real' scientific knowledge is also self-legitimating so is itself merely another 'narrative' (McLennan 1992: 332). Lyotard argues that scientific thought and knowledge has political and philosophical agendas and is therefore value laden and not totally objective (Harrington 2005). An ideal that underpinned the beginnings of Enlightenment thought was that the attainment of absolute knowledge for all would mean the attainment of absolute freedom for all (Calhoun, Rojek and Turner 2006).

It was thought that knowledge was the key to breaking down power structures that had existed during the domination of premodern narratives. Lyotard pointed to the contradictions within modern scientific bodies of knowledge and argued that they are themselves still made up of hierarchical power structures. Just as in earlier society narratives served to determine who had the right to speak and who did not, Lyotard states that this is still the case. Hierarchies still operate and serve to give the decision making elite the power to decide what gets defined as legitimate knowledge.

> Countless scientists have seen their "move" ignored or repressed, sometimes for decades, because it too abruptly destabilized the accepted positions, not only in the university and scientific hierarchy, but also in the problematic.
>
> (Lyotard, 1984: 63)

Lyotard argues therefore that science can no longer be seen as a unified body working towards the emancipation of humanity (Connor 1989: 31). His ideas can be linked to those of Nietzsche who believes that 'truth' and 'knowledge' are merely a 'matter of conventions that falsify and dissimulate to promote human survival' (Hollinger 1994: 105). Rather than being a grand quest for universal laws, Lyotard argues that knowledge is sought in order to keep human society functioning efficiently (Steuerman, 1992: 108). The assertion by science that it is constantly objectively striving for truth and progress is called into question by the fact that the search for knowledge is inextricably linked to achieving economic growth in society (McLennan 1992: 332). That is not to say that these two things are incompatible but rather to question if the search for truth and progress can ever be totally objective when there is a financial incentive. Lyotard argues that those involved in striving for progress:

> 'allocate our lives for the growth of power. In matters of social justice and of scientific truth alike, the legitimation of that power is based on its optimizing the system's performance – efficiency.
>
> (Lyotard, 1984: xxiv)

Since, as Lyotard (1984) points out, the legitimation of science can be called into question, knowledge as a unified, overarching metanarrative (such as the Enlightenment narrative that knowledge = liberation) as was key in modernity, breaks down into a wide range of 'micronarratives' (Lyotard, 1984: xxiv). Each separate specialism has a different discourse and play different 'language games' in an attempt to gain accreditation from their specific audience (Harrington 2005). This to Lyotard is what ultimately characterises the 'Postmodern Condition': *'Simplifying to the extreme, I define postmodern as incredulity toward metanarratives'* (Lyotard 1984: xxiv).

Lyotard's rejection of grand narratives obviously has strong implications for social theory. By rejecting the belief in the ability to universalise, philosophical thought loses its authority to make any suggestions as to what action can be taken in order to make changes in society (Best and Kellner 2001). In response to the criticisms of the Enlightenment made by Lyotard, Habermas wishes to consolidate the 'project of modernity' and further argues that we should not completely abandon the possibility of a rational pursuit of truth (Sibeon 2004). He defends modernity and argues that what is needed is more philosophical discussion, not less (Sibeon 2004). Habermas states that through the use of communicative action, language and rational dialogue, the Enlightenment aims of truth, justice and freedom are still attainable alongside social consensus (Butler 2002). However, Lyotard argues that Habermas ignores the fact that communication cannot simply take the form of consensual, rational dialogue, it will always take place in the context of power struggles (Bertens 1995: 128).

Therefore, rather than holding on to the ideals of Enlightenment scientific thought, Lyotard might suggest that we attempt to 'restructure' social theory in a postmodern vein so that we might find a democratic, pluralistic solution (Seidman 2004). He argued that whilst grand narratives, such as the Marxian narrative of class conflict, were well intentioned and essential to modernist sociology, to continue to utilise such concepts failed adequately to challenge the hierarchical structures in society (even though such hierarchies have become less steep over time) and therefore continued to marginalize and repress issues of difference.

> Postmodern science abandons absolute standards, universal categories and grand theories in favour of local, contextualized, and pragmatic conceptual strategies.
>
> (Seidman 1998: 207)

Seidman outlines the postmodern idea that the splintering of metanarratives has occurred to such an extent that society has become decentered. That is to say that there is no longer a common unifying culture in existence. (Seidman 2004). Individuals experience their lives at constantly shifting intersections of different discourses and language games. Instances of oppression, therefore, occur in many different contexts as individuals constantly construct, deconstruct and reconstruct themselves in terms of these 'fractured identities' (Bradley 1996: 211).

However, where this could not be explained utilising modern grand narratives which centred on one concept such as class conflict, Lyotard argues that a postmodern analysis does offer a way of explaining issues of multiplicity and difference (Lyotard 1984: 81). Employing Wittgenstein's idea of language games, Lyotard calls for the abandonment of any search for 'hidden' meanings along the lines, for example, of a depth hermeneutic approach to the study of social life, in favour of the 'play' of language games about which he makes three observations. First, the rules of such games do not contain their own legitimation, but are part of a contract between players who do not, necessarily, formulate those rules. Second, in the absence of rules, there is no game. In this way, an alteration in rules leads to an alteration in the nature of the game itself. Third, it follows that 'utterances' should be seen as moves within a game.

The third point may be given particular emphasis if we consider it in relation to Habermas's ideas, for 'to speak is to fight, in the sense of playing, and speech acts fall within the domain of a general agonistics' (Lyotard 1984: 10). This is not to suggest that people then 'play to win', simply that there is joy and pleasure to be gained in 'the endless invention of turns of phrase, of words and meanings, the process behind the evolution of language on the level of parole' (1984: 10). This pleasure, however, is still derived from the achievement of success over one's adversary. Underlying this process being the rules of the language game which provide for an 'observable common bond'.

Lyotard divides the normal methodological representation of the 'common bond' into two approaches. First, the functional holism of Parsons, and second, the conflict model of Marxism. The former is seen to chart an optimism in the faith of the system to adjust itself to the demands of stabilization in the age of globalization. The latter, on the other hand, as it evolved over time through the Frankfurt School, came to represent less of a theoretical model of critique based upon an analysis of social divisions, but acquired the status of 'utopia' or 'hope'. In the face of such contrasts we appear to be left with the choice between a technocratic positivism which feeds the system and a critical, hermeneutic or reflexive social theory (Lyotard 1984: 14). The point being, however, that this choice is no longer relevant to the society in which we live. In the postmodern condition, the functions of regulation and reproduction are no longer the province of particular groups or institutions. Instead, it is machines who hold information. Whilst those who have access to that information will continue to make decisions:

> What is new in all of this is that the old poles of attraction represented by nation-states, parties, professions, institutions, and historical traditions are losing their attraction. And it does not look as though they will be replaced, at least not on their former scale.
>
> (Lyotard 1984: 14)

The implication is that if these sources of the social bond are absent from our current social condition, then they should be sought elsewhere: for example, in social origins, or in some authentic history that requires de-coding. However, it cannot be uncovered in a single lifeworld and then used to explain or understand other lifeworlds. Given the necessity for the activation of differences within language games, there is no principle or general rules that enable us to establish meaningful connections. Indeed, the very attempt renders violence to heterogeneity.

In contrast to these strategies, we should understand the ways in which we are part of narratives. It is these language games that position 'the person who asks, as well as the addressee and the referent asked about: it is already the social bond' (Lyotard 1984: 15). In the age of information, these narratives flow through society and participation does not require justification in terms of reference to some grand, legitimizing narrative. As such, any person may occupy the place of the 'speaker'. These narratives are part of a communities relationship to itself and in their transmission we find 'the set of pragmatic rules that constitutes the social bond' (Lyotard 1984: 21).

As Baudrillard said of exponential cities, the meaning of narratives is not derived in relation to a past, but in the immediate act of their telling. This does not require 'science'. Nor does it require legitimation through allusions to humanism: that is, a referent to the autonomy of the will based in reason. The self is now part of the narrative society. As with the systems of rational coordination, the autonomous individual is part of a bygone era that is remembered only in the narratives of utopian radicals. In the dissemination of numerous language games without a common thread: 'The social subject itself seems to dissolve' (Lyotard 1984: 40). Systemic and critical modernism is thereby left in its wake.

As for Baudrillard, Lyotard argues that modernity contained the seeds of its own destruction. The desire for freedom, which is said to come with the greater accumulation of knowledge, is seen as symptomatic of Nietzsche's will to power. Scientific narratives, in the service of reason and progress, are no longer legitimate. Instead, they are part of a self-serving game, symptomatic of the forward march of de-legitimation. What is now at stake is the optimal relationship between inputs and outputs. This is not knowledge in the service of truth, but 'performativity'. Therefore, when it comes to research:

> The State and/or company must abandon the idealist and humanist narratives of legitimation in order to justify the new goal: in the discourse of today's backers of research, the only credible goal is power. Scientists, technicians, and instruments are purchased not to find truth, but to augment power.
>
> (Lyotard 1984: 46)

This system may not be overcome through the allusion to meta-narratives of explanation. It exists alongside narratives whose heterogeneity and dispersion is a cause for celebration. Given this, social theory should tell alternative stories in order to activate these differences and place its 'nostalgia' for 'the whole and the one' finally to rest.

REFERENCES

Adam, B. (1995) *Timewatch: The Social Analysis of Time.* Cambridge: Polity.

Adorno, T. and Horkheimer, M. (1979) *Dialectic of Enlightenment.* Originally published in 1944. Translated by Cumming, J. London: Verso.

Adorno, T., Frankel-Brunswik, E., Levinson, D. and Sanford, R. (1950) *The Authoritarian Personality.* New York: Harper & Row.

Adorno, T., Albert, H., Dahrendorf, R., Habermas, J., Pilot, H. and Popper, K. R. (1976) *The Positivist Dispute in German Sociology.* Translated by Adey, G. and Frisby, D. London: Heinemann.

Albrow, M. (1987) 'The Application of the Weberian Concept of Rationalization to Contemporary Conditions'. In Lash, S. and Whimster, S. (eds).

Albrow, M. (1990) *Max Weber's Construction of Social Theory.* London: Macmillan.

Alexander, J. C. (1984) 'The Parsons Revival in German Sociology'. *Sociological Theory* 2: 394-412.

Alexander, J. C. (ed.) (1985) *Neofunctionalism.* Beverly Hills, Calif. and London: Sage.

Alexander, J. C. (1990) 'The Centrality of the Classics'. In Giddens, A. and Turner, J. (eds).

Alexander, J. C. (1991) 'Habermas and Critical Theory: Beyond the Marxian Dilemma?' In Honneth, A. and Joas, H. (eds).

Alexander, J. C. (1995) *Fin de Siicle Social Theory: Relativism, Reduction and the Problem of Reason.* London: Verso.

Almquist, E. M. (1979) 'Black Women and the Pursuit of Equality'. In Freeman, J. (ed.) .

Althusser, L. (1969) *For Marx.* Translated by Brewster, B. Harmondsworth: Penguin.

Anderson, R., Hughes, J. and Sharrock, W. (1986) *Philosophy and the Human Sciences.* London: Routledge.

Archer, M. S. (1989) *Culture and Agency: The Place of Culture in Social Theory.* Cambridge: Cambridge University Press.

Archer, M. S. (1990) 'Human Agency and Social Structure: A Critique of Giddens'. In Clark, J., Modgil, C. and Modgil, S. (eds).

Armstrong, T. J. (ed.) (1992) *Michel Foucault: Philosopher.* Translated by Armstrong, T. J. London: Harvester Wheatsheaf.

Aronowitz, S. (1992) 'The Tensions of Critical Theory: Is Negative Dialectics All There Is?' In Seidman, S. and Wagner, D. (eds).

Assister, A. (1983) 'Did Man Make Language?' *Radical Philosophy* 34: 25-29.

Atkinson, J. M. (1978) *Discovering Suicide: Studies in the Social Organisation of Sudden Death.* London: Macmillan.

Atkinson, J. M. and Heritage, J. C. (eds) (1984) *Structures of Social Action: Studies in Conversation Analysis.* Cambridge: Cambridge University Press.

Atkinson, P. (1988) 'Ethnomethodology: A Critical Review'. *Annual Review of Sociology* 14: 441465.

Austin, J. (1976) *How to Do Things with Words.* Oxford: Oxford University Press.

Barber, W. J. (1981) *A History of Economic Thought.* Harmondsworth: Penguin.

Barrett, M. (1988) *Women's Oppression Today: The MarxistlFeminist Encounter.* Revised edn. London: Verso.

Barrett, M. and Phillips, A. (eds) (1992) *Destabilizing Theory: Contemporary Feminist Debates.* Cambridge: Polity.

Baudrillard, J. (1981) *For a Critique of the Political Economy of the Sign.* Originally published in 1972. St Louis, Miss.: Telos.

Baudrillard, J. (1983) *Simulations.* New York: Semiotext(e).

Baudrillard, J. (1993) 'Hyperreal America'. Translated by Macey, D. *Economy and Society* 22(2): 243-252.

Bauman, Z. (1978) *Hermeneutics and Social Science: Approaches to Understanding.* London: Hutchinson.

Bauman, Z. (1987) *Legislators and Interpreters: On Modernity, Post-Modernity and Intellectuals.* Cambridge: Polity.

Bauman, Z. (1989) *Modernity and the Holocaust.* Cambridge: Polity.

Bauman, Z. (1991) *Modernity and Ambivalence.* Cambridge: Polity.

Bauman, Z. (1992) *Intimations of Postmodernity.* London: Routledge.
Beck, U. (1992) *Risk Society: Towards a New Modernity.* London: Sage.
Beck, U., Giddens, A. and Lash, S. (1994) *Reflexive Modernization: Politics, Tradition and Aesthetics in the Modern Social Order.* Cambridge: Polity.
Becker, H. (1963) *Outsiders: Studies in the Sociology of Deviance.* New York: Free Press.
Beechey, V. (1986) 'Women and Employment in Contemporary Britain'. In Beechey, V. and Whitelegg, E. (eds).
Beechey, V. and Whitelegg, E. (eds) (1986) *Women in Britain Today.* Milton Keynes: Open University Press.
Beetham, D. (1974) *Max Weber and the Theory of Modern Politics.* London: Allen & Unwin.
Bell, D. (1960) *The End of Idelogy.* New York: Collier.
Beloff, M. (1954) *The Age of Absolutism: 1660-1 81 5.* London: Hutchinson.
Bendix, R. (1977) *Max Weber: An Intellectual Portrait.* Berkeley, Calif.: University of California Press.
Benhabib, S. (1992) *Situating the Self: Gender, Community and Postmodernism in Contemporary Ethics.* Cambridge: Polity.
Benjamin, A. (ed.) (1991) *The Problems ofModernity: Adornoand Benjamin.* London: Routledge.
Benson, D. (1974) 'Critical Note: A Revolution in Sociology'. *Sociology* 8: 125-129.
Benton, T. (1977) *Philosophical Foundations of the Three Sociologies.* London: Routledge & Kegan Paul.
Berger, P. L. and Luckmann, T. (1 967) *The Social Construction of Reality: A Treatise in the Sociology of Knowledge.* New York: Anchor.
Berman, M. (1988) *All That is Solid Melts into Air: The Experience of Modernity.* 2nd edn. London: Verso.
Berman, M. (1993) 'Why Modernism Still Matters'. In Lash, S. and Friedman, J. (eds).
Bernauer, J. and Rasmussen, D. (eds) (1991) *The Final Foucault.* Cambridge, Mass.: MIT Press.
Bernstein, R. (1983) *Beyond Objectivism and Relativism: Science, Hermeneutics and Praxis.* Oxford: Basil Blackwell.
Bernstein, R. (1985a) 'Introduction'. In Bernstein, R. (ed.).
Bernstein, R. (ed.) (1985b) *Habermas and Modernity.* Cambridge: Polity.
Bershady, H. J. (1991) 'Practice against Theory in American Sociology: An Exercise in the Sociology of Knowledge'. In Robertson, R. and Turner, B. S. (eds)

Best, S. (1 994) 'The Commodification of Reality and the Reality of Commodification:
Baudrillard, Debord, and Postmodern Theory'. In Kellner, D. (ed.).
Bhaskar, R. (1993) *Dialectic: The Pulse of Freedom.* London: Verso.
Bierstedt, R. (1 979) 'Sociological Thought in the Eighteenth Century'. In Bottomore, T. and Nisbet, R. (eds).
Billig, M. (1994) 'Sod Baudrillard! Or Ideology Critique in Disney World'. In Simons, H. W. and Billig, M. (eds).
Bittner, E. (1974) 'The Concept of Organization'. Originally published in 1965. Reprinted in Turner, R. (ed.).
Blackman, H. J. (ed.) (1967) *Objections to Humanism.* Harmondsworth: Penguin.
Blackman, H. J. (1968) *Humanism.* Harmondsworth: Penguin.
Bloor, D. (1983) *Wittgenstein: A Social Theory of Knowledge.* London: Macmillan.
Blumer, H. (1937) 'Social Psychology'. In Schmidt, E. P. (ed.).
Blumer, H. (1969) *Symbolic Interactionism: Perspective and Method.* Englewood Cliffs, NJ: Prentice-Hall.
Blumer, H. (1972) 'Society as Symbolic Interaction'. In Rose, A. (ed.).
Blumer, H. (1983) 'Going Astray with a Logical Scheme'. Reprinted in Plummer, K. (ed.) (1991 a).
Boden, D. and Zimmerman, D. H. (eds) (1993) *Talk and Social Structure: Studies in Ethnomethodology and Conversation Analysis.* Cambridge: Polity.
Bosanquet, T. (1967) 'Auguste Comte and the Positive Philosophers'. In Hearnshaw, F. J. *C.* (ed.).
Bottomore, T. (1979) 'Marxism and Sociology'. In Bottomore, T. and Nisbet, R. (eds).
Bottomore, T. and Nisbet, R. (1979a) 'Structuralism'. In Bottomore, T. and Nisbet, R. (eds) .
Bottomore, T. and Nisbet, R. (eds) (1979b) A *History of Sociological Analysis.* London: Heinemann.
Bourdieu, P. (1 962) *The Algerians.* Boston, Mass.: Beacon.
Bourdieu, P. (1977) Outline of a Theory of Practice. Translated by Nice, R. Cambridge: Cambridge University Press.
Bourdieu, P. (I98 1) 'Men and Machines'. In Knorr-Cetina, K. and Cicourel, A. (eds).
Bourdieu, P. (1984) Distinction: A Social Critique of the Judgement of Taste. Translated by Nice, R. London: Routledge & Kegan Paul.

Bourdieu, P. (1988) Homo Academicus. Translated by Collier, P. Cambridge: Polity.

Bourdieu, P. (1989) 'Social Space and Symbolic Power'. Sociological Theory 7(1): 14-25.

Bourdieu, P. (1990) In Other Words: Essays Towards a Reflexive Sociology. Translated by Adamson, M. Cambridge: Polity.

Bourdieu, P. (1991a) 'Meanwhile, I Have Come to Know All the Diseases of Sociological Understanding: An Interview'. Conducted by Krais, B. In Bourdieu, P. Chamboredon, J.-C. and Passeron, J.-C.

Bourdieu, P. (1 991 b) The Political Ontology of Martin Heidegger. Translated by Collier, P. Cambridge: Polity.

Bourdieu, P. (1992a) The Logic of Practice. Translated by Nice, R. Originally published in 1980 as Le Sens pratique. Cambridge: Polity.

Bourdieu, P. (1992b) Language and Symbolic Power. Edited and introduced by Thompson, J. B. Translated by Raymond, G. and Adamson, M. Cambridge: Polity.

Bourdieu, P. (1993a) 'Concluding Remarks: For a Sociogenetic Understanding of Intellectual Works'. In Calhoun, C., LiPuma, E. and Postone, M. (eds).

Bourdieu, P. (1993b) Sociology in Question. Translated by Nice, R. London: Sage.

Bourdieu, P. (1995) The Rules of Art: Genesis and Structure of the Literary Field. Translated by Emanuel, S. Cambridge: Polity.

Bourdieu, P. and Passeron, J.-C. (1990) Reproduction in Education, Society and Culture. 2nd edn. London: Sage.

Bourdieu, P., Chamboredon, J.-C. and Passeron, J.-C. (1991) The Craft of Sociology: Epistemological Preliminaries. Edited by Krais, B. Translated by Nice, R. New York: Walter de Gruyter.

Bourdieu, P. and Wacquant, L. J. (1992) An Invitation to Reflexive Sociology. Cambridge: Polity.

Bourdieu, P, Passeron, J.-C. and de Saint Martin, M., with contributions by Baudelot, C. and Vincent, G. (1994) Academic Discourse: Linguistic Misunderstanding and Professorial Power. Translated by Teese, R. Cambridge: Polity.

Boyne, R. (1991) 'Power-Knowledge and Social Theory: The Systematic Misrepresentation of Contemporary French Social Theory in the Work of Anthony Giddens'. In Bryant, C. and Jary, D. (eds).

Brenner, J. and Ramas, M. (1990) 'Rethinking Women's Oppression'. In Lovell, T. (ed.).

Bronner, S. (1994) *Of* Critical Theory and its Theorists. Oxford: Blackwell.

Bronner, S. and Kellner, D. (eds) (1989) Critical Theory and Society: A Reader. London: Routledge.

Bronowski, J. and Mazlish, B. (1970) The Western Intellectual Tradition. Harmondsworth: Penguin.

Brown, P. and Jordanova, L. J. (1981) 'Oppressive Dichotomies: The Nature/Culture Debate'. In Cambridge Women's Studies Group (eds).

Brubaker, R. (1985) 'Rethinking Classical Theory: The Sociological Vision of Pierre Bourdieu'. Theory and Society 14(6): 745-775.

Brunkhorst, H. (1992) 'Culture and Bourgeois Society: The Unity of Reason in a Divided Society'. In Honneth, A., McCarthy, M., Offe, C. and Wellmer, A. (eds) (1 992)

Brunt, R. and Rowan, C. (eds) (1982) *Feminism, Culture and Politics.* London: Lawrence & Wishart.

Bryant, C. and Jary, D. (eds) (1991) *Giddens Theory of Structuration: A Critical Appreciation.* London: Routledge.

Bulmer, M. (1984) *The Chicago School of Sociology: Institutionalization, Diversity, and the Rise of Sociological Research.* Chicago: University of Chicago Press.

Burchell, G., Gordon, C. and Miller, P. (eds) (1991) *The Foucault Effect: Studies in Govermentality.* London: Harvester Wheatsheaf.

Burke, K. (1969) *A Grammar of Motives.* Originally published in 1945. Berkeley, Calif.: University of California Press.

Burke, P. (1980) *Sociology and Histo y.* London: Allen & Unwin.

Burns, T. (1992) *Erving Goffman.* London: Routledge.

Button, G. (1991a) 'Introduction: Ethnomethodology and the Foundational Respecification of the Human Sciences'. In Button, G. (ed.).

Button, G. (ed) (1991b) *Ethnomethodology and the Human Sciences.* Cambridge: Cambridge University Press.

Button, G. and Casey, N. (1984) 'Generating Topic: The Use of Topic Initial Elicitors'. In Atkinson, J. M. and Heritage, J. (eds).

Bynner, J. and Stribley, K. (eds) (1979) *Social Research: Principles and Procedures.* Milton Keynes: Open University Press.

Calhoun, C. (ed.) (1994) *Social Theory and the Politics of ldentity.* Oxford: Blackwell.

Calhoun, C. (1995) *Critical Social Theory: Culture, History, and the Challenge of Difference.* Oxford: Blackwell.

Calhoun, C., LiPuma, E. and Postone, M. (eds) (1993) *Bourdieu: Critical Perspectives.* Cambridge: Polity.

Callinicos, A. (1990) *Against Postmodernism: A Marxist Critique.* Cambridge: Polity.
Callinicos, A. (1 993) 'What is Living and What is Dead in the Philosophy of Althusser?' In Kaplan, E. and Sprinker, M. (eds).
Cambridge Women's Studies Group (eds) (198 1) *Women in Society: Interdisciplinary Essays.* London: Virago.
Chappell, V. C. (ed.) (1968) *Hume.* London: Macmillan.
Chodorow, N. J. (1978) *The Reproduction of Mothering: Psychoanalysis and the Sociology of Gender.* Berkeley, Calif.: University of California Press.
Chodorow, N. J. (1986) 'Towards a Relational Individualism: The Mediation of Self through Psychoanalysis'. In Heller, T., Sosna, M. and Wellbery, D. with Davidson, A., Swidler, A. and Watt, I. (eds).
Chomsky, N. (1980) *Rules and Representation.* Oxford: Blackwell.
Cicourel, A. V. (1964) *Method and Measurement in Sociology.* London: Macmillan.
Cicourel, A. V. (1973) *Cognitive Sociology: Language and Meaning in Social Interaction.* Harmondsworth: Penguin.
Cicourel, A. V. (1 976) *The Social Organisation of Juvenile Justice.* 2nd edn. London: Heinemann.
Cicourel, A. V. (1981) 'Notes on the Integration of Micro- and Macro-Levels of Analysis'. In Knorr-Cetina, K. and Cicourel, A. (eds).
Cicourel, A. V. (1993) 'Aspects of Structural and Processual Theories of Knowledge'. In Calhoun, C., LiPuma, E. and Postone, M. (eds).
Clammer, J. (1976) 'Wittgensteinianism and the Social Sciences'. *Sociological Review* 24(4) : 775-79 1.
Clark, J., Modgil, C. and Modgil, S. (eds) (1990) *Anthony Giddens: Consensus and Controversy.* Brighton: Falmer.
Clark, S. (1991) 'The *Annales* Historians'. In Skinner, Q. (ed.).
Cleaver, H. (1979) *Reading 'Capital' Politically.* Brighton: Harvester.
Cockburn, C. (1 990) 'The Material of Male Power'. In Lovell, T. (ed.).
Cohen, G. A. (1984) *Karl Marx's Theory of History: A Defence.* Oxford: Oxford [Jniversity Press.
Cohen, I. (1989) *Structuration Theory: Anthony Giddens and the Constitution of Social Life.* London: Macmillan.
Coleman, J. S. (1990) *Foundations of Social Theory.* Cambridge, Mass.: Belknap.
Collins, R. (1981) 'Micro-Translation as a Theory Building Strategy'. In Knorr-Cetina, K. and Cicourel, A. (eds).
Collins, R. (ed.) (1984) *Sociological Theory.* San Francisco: Josey Bass.

Cooke, M. (1994) *Language and Reason: A Study of Habermas's Pragmatics.* Cambridge, Mass.: MIT Press.
Cooley, C. H. (1902) *Human Nature and the Social Order.* New York: Free Press.
Coser, L. (1979) 'American Trends'. In Bottomore, T. and Nisbet, R. (eds).
Coulter, J. (1979) *The Social Construction of Mind: Studies in Ethnomethodology and Linguistic Philosophy.* London: Macmillan.
Coulter, J. (1983) *Rethinking Cognitive Theory.* London: Macmillan.
Coulter, J. (ed.) (1990) *Ethnomethodological Sociology.* Aldershot: Edward Elgar.
Coward, R. (1 982) 'Sexual Politics and Psychoanalysis'. In Brunt, R. and Rowan, C. (eds) .
Craib, I. (1992) *Anthony Giddens.* London: Routledge.
Crook, S., Pakulski, J. and Waters, M. (1992) *Postmodernization: Change in Advanced Society.* London: Sage.
Crosland, M. (1992) *Simone de Beauvoir: The Woman and her Work.* London: Heinemann.
Crowley, H. and Himmelweit, S. (eds) (1992) *Knowing Women: Feminism and Knowledge.* Cambridge: Polity.
Dahrendorf, R. (1959) *Class and Class Conflict in Industrial Society.* London: Routledge & Kegan Paul.
Davis, K., Leijenaar, M. and Oldersma, J. (eds) (1991) *The Gender ofpower.* London: Sage.
Dawe, A. (1970) 'The Two Sociologies'. *British Journal of Sociology.* 21: 207-218.
Dean, M. (1994) *Critical and Effective Histories: Foucault's Methods and Historical Sociology.* London: Routledge.
de Beauvoir, S. (1972) *The Second Sex.* Originally published in 1949. Harmondsworth: Penguin.
de Lauretis, T. (1990) 'Upping the Anti *[sic]* in Feminist Theory'. In Hirsch, M. and Fox Keller, E. (eds).
Delmar, R. (1976) 'Looking Again at Engels' "Origin of the Family, Private Property and the State" '. In Oakley, A. and Mitchell, J. (eds).
Delphy, C. and Leonard, D. (1992) *Familiar Exploitation: A New Analysis of Marriage in Contemporary Western Society.* Cambridge: Polity.
Denzin, N. K. (1974) 'Symbolic Interactionism and Ethnomethodology7. In Douglas, J. D. (ed.).
Denzin, N. K. (1 992) *Symbolic Interactionism and Cultural Studies: The Politics of Interpretation.* Oxford: Blackwell.

Derrida, J. (1978) *Writing and Difference*. London: Routledge.
Dewey, J. (1896) 'The Reflexive Arc Concept in Psychology'. *Psychological Review* 3: 357-370.
Dews, P. (1989) 'The Return of the Subject in the Late Foucault'. *Radical Philosophy* 51: 37-41.
Dews, P. (ed.) (1992) *Autonomy and Solidarity: Interviews with Jiirgen Habermas*. Revised edn. London: Verso.
Diamond, I. and Quinby, L. (eds) (1988) *Feminism and Foucault: Reflections on Resistance*. Boston, Mass.: Northeastern University Press.
Ditton, J. (ed.) (1980) *The View from Goffman*. London: Macmillan.
Docherty, T. (ed.) (1 993) *Postmodernism: A Reader*. London: Harvester Wheatsheaf.
Doney, W. (ed.) (1967) *Descartes: A Collection of Critical Essays*. London: Macmillan.
Dore, R. P. (1973) 'Function and Cause'. In Ryan, A. (ed.).
Douglas, J. D. (ed.) (1974) *Understanding Everyday Life: Towards the Reconstruction of Sociological Knowledge*. London: Routledge & Kegan Paul.
Douglas, M. (1970) *Purity and Danger: An Analysis of the Concepts of Pollution and Taboo*. New York: Pantheon.
Douglas, M. (1978) *Implicit Meanings: Essays in Anthropology*. London: Routledge & Kegan Paul.
Drew, P. and Wootton, A. (eds) (1988) *Eruing Goffman: Exploring the Interaction Order*. Cambridge: Polity.
Dreyfus, H. and Rabinow, P. (1982) *Michel Foucault: Beyond Structuralism and Hermeneutics*. Chicago: University of Chicago Press.
Dreyfus, H. and Rabinow, P. (1 993) 'Can There be a Science of Existential Structure and Social Meaning?' In Calhoun, C., LiPuma, E. and Postone, M. (eds).
Du Bois, W. (1989) *The Souls of Black Folk*. Originally published in 1903. New York: Bantam.
Durkheim, E. (1964) *The Division of Labour in Society*. Originally published in 1893. Translated by Simpson, G. New York: Free Press.
Durkheim, E. (1983) *Pragmatism and Sociology*. Originally published in 1955. Edited by Allcock, J. B. Translated by Whitehouse, J. C. Cambridge: Cambridge University Press.
Durkheim, E. (1992) 'Review Article on F. S. Merlino's "Formes et essences du socialisme" '. Originally published in 1899. In Gane, M. (ed.).
Dworkin, A. (1987) *Intercourse*. London: Secker & Warburg.

Easthope, A. and McGowan, K. (eds) (1992) *A Critical and Cultural Theory Reader.* Milton Keynes: Open University Press.

Eco, U. (1993) 'The City of Robots'. In Docherty, T. (ed.).

Elders, F. (ed.) (1974) *Reflexive Waters: The Basic Concerns of Mankind.* London: Souvenir.

Elias, N. (1991) *The Symbol Theory.* London: Sage.

Elliott, A. (1992) *Social Theory and Psychoanalysis in Transition: Self and Society from Freud to Kristeva.* Oxford: Blackwell.

Engels, F. (1942) *The Origin of the Family, Private Property and the State.* Originally published in 1884. New York: International Publishers.

Evans, M. (ed.) (1982) *The Woman Question: Readings on the Subordination of Women.* London: Fontana.

Faberman, H. A. (1985) 'The Foundations of Symbolic Interactionism: James, Cooley and Mead'. Reprinted in Plummer, K. (ed.) (1991a).

Fairclough, N. (1989) *Language and Power.* London: Longman.

Farganis, S. (1994) *Situating Feminism: From Thought to Action.* London: Sage.

Featherstone, M. (ed.) (1 991) *Consumer Culture and Postmodernism.* London: Sage.

Featherstone, M. and Turner, B. (1995) 'Body and Society: An Introduction'. *Body and Society* l(1)1: -12.

Figes, E. (1978) *Patriarchal Attitudes: Women in Society.* Originally published in 1970. London: Virago.

Firestone, S. (1988) *The Dialectic of Sex: The Case for Feminist Revolution.* Originally published 1970. London: Women's Press.

Fisher, B. M. and Strauss, A. L. (1978) 'The Chicago Tradition and Social Change: Thomas, Park and their Successors'. Reprinted in Plummer, K. (ed.) (1991a).

Fisher, B. M. and Strauss, A. L. (1979) 'Interactionism'. In Bottomore, T. and Nisbet, R. (eds).

Foucault, M. (1979) *The History of Sexuality, Volume* 1: *An Introduction.* Translated by Hurley, R. Harmondsworth: Penguin.

Foucault, M. (1980) *PowerlKnowledge, Selected Interviews and Other Writings* 1972-1 977. Edited by Gordon, C. Brighton: Harvester.

Foucault, M. (1982) 'The Subject and Power'. In Dreyfus, H. and Rabinow, P. (eds).

Foucault, M. (1984) *The Foucault Reader.* Edited by Rabinow, P. Harmondsworth: Penguin.

Foucault, M. (1988a) 'The Political Technology of Individuals'. In Martin, L. H., Gutman, H. and Hutton, P. H. (eds).

Foucault, M. (1988b) 'Technologies of the Self'. In Martin, L. H., Gutman, H. and Hutton, P. H. (eds).

Foucault, M. (1988~'T) ruth, Power, Self: An Interview'. In Martin, L. H., Gutman, H. and Hutton, P. H. (eds).

Foucault, M. (1989) *The Archaeology of Knowledge.* Originally published in 1969. London: Routledge.

Foucault, M. (1991a) *Remarks on Marx: Conversations with Duccio Trombadori.* Translated by Goldstein, R. J. and Cascaito, J. New York: Semiotext(e).

Foucault, M. (1991b) 'The Ethic of Care for the Self as a Practice of Freedom: An interview with Fornet-Betancourt, R., Becker, H. and Gomez-Miiller, A. Translated by Gauthier Snr, J. D. In Bernauer, J. and Rasmussen, D. (eds).

Foucault, M. (1991~')Q uestions of Method'. In Burchell, G., Gordon, C. and Miller, P. (eds).

Foucault, M. (1991d) *Discipline and Punish: The Birth of the Prison.* Originally published in 1977. Translated by Sheridan, A. Harmondsworth: Penguin.

Foucault, M. (1992) *The Order of Things: An Archaeology of the Human Sciences.* Originally published in 1970. London: Routledge.

Fraser, N. (1989) *Unruly Practices: Power, Discourse and Gender in Contemporary Social Theory.* Cambridge: Polity.

Freeman, J. (ed.) (1979) *Women: A Feminist Perspective.* 2nd edn. Palo Alto, Calif.: May field.

Freud, S. (1938) *Psychopathology of Everday Life.* Harmondsworth: Penguin.

Freud, S. (1969) *Civilization and its Discontents.* Translated by Riviere, J. Edited by Strachey, J. London: Hogarth.

Freund, J. (1979) 'German Sociology in the Time of Max Weber'. In Bottomore, T. and Nisbet, R. (eds).

Freundlieb, D. (1994) 'Foucault's Theory of Discourse and Human Agency'. In Jones, C. and Porter, R. (eds).

Friedan, B. (1965) *The Feminine Mystique.* Harmondsworth: Penguin.

Frisby, D. (1991) 'The Aesthetics of Modern Life: Simmel's Interpretation'. In Featherstone, M. (ed.).

Frisby, D. (1992a) *Sociological Impressionism: A Reassessment of Georg Simmel's Social Theory.* 2nd edn. London: Routledge.

Frisby, D. (1992b) *Simmel and Since: Essays on Georg Simmel's Social Theory.* London: Routledge.
Fromm, E. (1969) *Escape from Freedom.* Originally published in 1941. New York: Avon.
Gadamer, H. G. (1975) *Truth and Method.* Originally published in 1960. London: Sheed & Ward.
Gane, M. (1990) 'Ironies of Postmodernism: Fate of Baudrillard's Fatalism'. *Economy and Society* 19(3): 314-334.
Gane, M. (1991) *Baudrillard: Critical and Fatal Theory.* London: Routledge.
Gane, M. (ed.) (1992) *The Radical Sociology of Durkheim and Mauss.* London: Routledge.
Gane, M. (ed.) (1993) *Baudrillard Live: Selected Interviews.* London: Routledge.
Garfinkel, H. (1967) *Studies in Ethnomethodology.* Englewood Cliffs, NJ: Prentice-Hall.
Garfinkel, H. (1974) 'The Origins of the Term "Ethnomethodology"'. Originally published in 1968. Reprinted in Turner, R. (ed.).
Garfinkel, H. (ed.) (1986) *Ethnomethodological Studies of Work.* London: Routledge & Kegan Paul.
Garfinkel, H. (1991) 'Respecification: Evidence for Locally Produced, Naturally Accountable Phenomena of Order, Logic, Reason, Meaning, Method, etc. In and as of the Essential Haecceity of Immortal Ordinary Society, (1) - An Announcement of Studies'. In Button, G. (ed.).
Garfinkel, H. and Sacks, H. (1986) 'On Formal Structures of Practical Actions'. In Garfinkel, H. (ed.).
Gartman, D. (1991) 'Culture as Class Symbolization or Mass Reification? A Critique of Bourdieu's *Distinction'. American lournal of Sociology* 97(2): 421-447.
Gellner, E. (1968) *Words and Things.* Penguin: Harmondsworth.
Gellner, E. (1974) 'The New Idealism: Cause and Meaning in the Social Sciences'. In Giddens, A. (ed.).
Gellner, E. (1988) *Culture, Identity and Politics.* Cambridge: Cambridge University Press.
Gellner, E. (1992) *Reason and Culture: The Historic Role of Rationality and Rationalism.* Oxford: Blackwell.
Gerth, H. and Mills, C. W. (eds) (1970) *From Max Weber: Essays in Sociology.* Originally published in 1948. London: Routledge & Kegan Paul.

Giddens, A. (1 971) *Capitalism and Modern Social Theory: An Analysis of the Writings of Marx, Durkheim and Max Weber.* Cambridge: Cambridge University Press.

Giddens, A. (1972a) *Politics and Sociology in the Thought of Max Weber.* London: Macmillan.

Giddens, A. (ed.) (1972b) *Emile Durkheim: Selected Writings.* Cambridge: Cambridge University Press.

Giddens, A. (ed.) (1 974) *Positivism and Sociology.* London: Heinemann.

Giddens, A. (1976) *New Rules of Sociological Method: A Positive Critique of Interpretive Sociologies.* London: Hutchinson.

Giddens, A. (1977) *Studies in Social and Political Theory.* London: Hutchinson.

Giddens, A. (1979) *Central Problems in Social Theory: Action, Structure and Contradiction in Social Analysis.* London: Macmillan.

Giddens, A. (198 la) *A Contemporary Critique of Historical Materialism: Volume 1: Power, Property and the State.* London: Macmillan.

Giddens, A. (1981 b) 'Agency, Institution, and Time-Space Analysis'. In Knorr-Cetina, K. and Cicourel, A. (eds).

Giddens, A. (1 982) *Profiles and Critiques in Social Theory.* Berkeley, Calif.: University of California Press.

Giddens, A. (1984) *The Constitution ofsociety: Outline ofthe Theory ofStructuration.* Cambridge: Polity.

Giddens, A. (1985) *The Nation-State and Violence: Volume 2 of A Contemporary Critique of Historical Materialism.* Cambridge: Polity.

Giddens, A. (1987) *Social Theory and Modern Sociology.* Cambridge: Polity.

Giddens, A. (1989) 'A Reply to my Critics'. In Held, D. and Thompson, J. B. (eds).

Giddens, A. (1990) *The Consequences of Modernity.* Cambridge: Polity.

Giddens, A. (1 99 la) 'Structuration Theory: Past, Present and Future'. In Bryant, C. and Jary, D. (eds) .

Giddens, A. (1991 b) *Modernity and Self-Identity.* Cambridge: Polity.

Giddens, A. (1 992) *The Transformation of Intimacy: Sexuality, Love and Eroticism in Modern Societies.* Cambridge: Polity.

Giddens, A. (1993) *New Rules of Sociological Method: A Positive Critique of Interpretive Sociologies.* 2nd edn. Cambridge: Polity.

Giddens, A. (1994) *Beyond Left and Right: The Future of Radical Politics.* Cambridge: Polity.

Giddens, A. and Turner, J. (eds) (1990) *Social Theory Today.* Cambridge: Polity.

Gjertsen, D. (1992) *Science and Philosophy: Past and Present.* Harmondsworth: Penguin.

Gliicksmann, A. (1992) 'Michel Foucault's Nihilism'. In Armstrong, T. (ed.).

Goffman, E. (1961) *Encounters: Two Studies in the Sociology of Interaction.* Harmondsworth: Penguin.

Goffman, E. (1968) *Asylums: Essays on the Social Situation of Mental Patients and Other Inmates.* Originally published in 1961. Harmondsworth: Penguin.

Goffman, E. (1971) *Relations in Public: Microstudies of the Public Order.* Harmondsworth: Penguin.

Goffman, E. (1972) *Interaction Ritual: Essays on Face-to-Face Behaviour.* Harmondsworth: Penguin.

Goffman, E. (1974) *Frame Analysis: An Essay on the Organization of Experience.* New York: Harper & Row.

Goffman, E. (1979) *Gender Advertisements.* London: Macmillan.

Goffman, E. (1981) *Forms of Talk.* Philadelphia, Pa: University of Pennsylvania Press.

Goffman, E. (1 983) 'The Interaction Order'. The 1982 American Sociological Association Presidential Address. Reprinted in Plummer, K. (ed.) (1991 b).

Goffman, E. (1984) *The Presentation of Self in Everyday Life.* Originally published in 1959. Harmondsworth: Penguin.

Goldthorpe, J. H. (1973) 'A Revolution in Sociology?' *Sociology* 7: 449462.

Goldthorpe, J. H. and Marshall, G. (1992) 'The Promising Future of Class Analysis'. *Sociology* 26(3) : 3 8 1400.

Gordon, C. (1987) 'The Soul of the Citizen: Max Weber and Michel Foucault on Rationality and Government'. In Lash, S. and Whimster, S. (eds).

Gouldner, A. (1 971) *The Coming Crisis in Western Sociology.* London: Heinemann.

Gramsci, A. (1971) *Prison Notebooks: Selections.* Translated by Hoare, Q. and Smith, G. N. New York: International Publishers.

Greer, G. (1970) *The Female Eunuch.* London: MacGibbon & Kee.

Gregson, N. (1989) 'On the (1r)relevance of Structuration Theory to Empirical Research'. In Held, D. and Thompson, J. B. (eds).

Griffiths, M. (1988) 'Feminism, Feelings and Philosophy'. In Griffiths, M. and Whitford, M. (eds).

Griffiths, M. and Whitford, M. (eds) (1988) *Feminist Perspectives in Philosophy.* London: Macmillan.

Grimshaw, J. (1989) 'Mary Wollstonecraft and the Tensions in Feminist Philosophy'. *Radical Philosophy* 52: 11-17.

Gross, E. (1992) 'What is Feminist Theory?' In Crowley, H. and Himmelweit, S. (eds).

Grosz, E. (1990) 'A Note on Essentialism and Difference'. In Gunew, S. (ed.).

Gunew, S. (ed.) (1990) *Feminist Knowledge: Critique and Construct.* London: Routledge.

Habermas, J. (1 975) *Legitimation Crisis.* Translated by McCarthy, T. Boston, Mass.: Beacon.

Habermas, J. (1981) 'Towards a Reconstruction of Historical Materialism'. In Knorr-Cetina, K. and Cicourel, A. (eds).

Habermas, J. (1984) *Theory of Communicative Action. Volume 1: Reason and the Rationalization of Society.* Translated by McCarthy, T. London: Heinemann.

Habermas, J. (1986) 'Taking Aim at the Heart of the Present'. In Hoy, D. C. (ed.).

Habermas, J. (1987) *Theory of Communicative Action. Volume 2: Lifeworld and System: A Critique of Functionalist Reason.* Translated by McCarthy, T. Cambridge: Polity.

Habermas, J. (1989a) *Knowledge and Human Interests.* Originally published in 1968. Translated by Shapiro, J. J. Cambridge: Polity.

Habermas, J. (1989b) *The Structural Transformation of the Public Sphere.* Translated by Berger, T. and Lawrence, F. Cambridge, Mass.: MIT Press.

Habermas, J. (1990) *On the Logic of the Social Sciences.* Originally published in1970. Translated by Nicholsen, S. W. and Stark, J. A. Cambridge: Polity.

Habermas, J. (1991) 'A Reply'. In Honneth, A. and Joas, H. (eds).

Habermas, J. (1992a) The Philosophical Discourse of Modernity: Twelve Lectures. Cambridge: Polity.

Habermas, J. (1992b) Postmetaphysical Thinking: Philosophical Essays. Translated by Hohengarten, W. M. Cambridge, Mass.: MIT Press.

Habermas, J. (1992~M) oral Consciousness and Communicative Action. Translated by Lenhardt, C. and Nicholsen, S. Introduction by McCarthy, T. Cambridge: Polity.

Habermas, J. (1993) Justification and Application: Remarks on Discourse Ethics. Translated by Cronin, C. Cambridge: Polity.

Habermas, J. (1994) The Past as Future. Interviewed by Haller, M. Translated and edited by Pensky, M. Cambridge: Polity.

Habermas, J. (1996) Between Facts and Norms: Contributions to a Discourse Theory of Law and Democracy. Cambridge: Polity.

Hage, J. (ed.) (1994) Formal Theory in Sociology: Opportunity or Pitfall? Albany, NY: State University of New York Press.

Hall, P. M. (1987) 'Interactionism and the Study of Social Organization'. Reprinted in Plummer, K. (ed.) (1991 b).

Hall, S. and Gieben, B. (eds) (1992) Formations of Modernity. Polity Press in association with the Open University. Cambridge: Polity.

Hamilton, P. (1983) Talcott Parsons. London: Tavistock.

Hamilton, P. (ed.) (1985) Readings from Talcott Parsons. London: Tavistock.

Hamilton, P. (1992) 'The Enlightenment and the Birth of Social Science'. In Hall, S. and Gieben, B. (eds).

Haraway, D. (1991) Simians, Cyborgs, and Women: The Reinvention of Nature. New York: Routledge.

Harding, S. (1986) The Science Question in Feminism. Milton Keynes: Open University Press.

Harding, S. (1991) Whose Science? Whose Knowledge? Thinking from Women's Lives. Milton Keynes: Open University Press.

Harding, S. and Hintikka, M. B. (1983a) 'Introduction'. In Harding, S. and Hintikka, M. B. (eds).

Harding, S. and Hintikka, M. B. (eds) (1983b) Discovering Reality: Feminist Perspectives on Epistemology, Metaphysics, Methodology, and Philosophy of Science. London: D. Reidel.

Hartsock, N. (1983) 'The Feminist Standpoint: Developing the Ground for a Specifically Feminist Historical Materialism'. In Harding, S. and Hintikka, M. B. (eds).

Harvey, D. (1990) The Condition of Postmodernity: An Enquiry into the Origins of Cultural Change. Oxford: Blackwell.

Hawthorn, G. (1976) Enlightenment and Despair: A History of Sociology. Cambridge: Cambridge University Press.

Hearn, J. and Morgan, D. (eds) (1990) Men, Masculinities and Social Theory. London: Unwin Hyman.

Hearnshaw, F. J. C. (ed.) (1967) The Social and Political Ideas of Some Representative Thinkers of the Age of Reaction and Reconstruction: *1815-1 865.* Originally published in 1932. London: Dawsons.

Hegel, G. W. F. (1967a) The Phenomenology of Mind. Translated by Baillie, J. B. Oxford: Oxford University Press.

Hegel, G. W. F. (1967b) *Hegel's Philosophy of Right.* Translated by Knox, T. M. Oxford: Oxford University Press.

Heilbron, J. (1995) *The Rise of Social Theory.* Translated by Gogol, S. Cambridge: Polity.

Hekman, S. (1990a) 'Hermeneutics and the Crisis of Social Theory: A Critique of Giddens's Epistemology'. In Clark, J., Modgil, C. and Modgil, S. (eds).

Hekman, S. (1990b) *Gender and Knowledge: Elements of a Postmodern Feminism.* Cambridge: Polity.

Held, D. (1990) *introduction to Critical Theory: Horkheimer to Habermas.* Cambridge: Polity.

Held, D. and Thompson, J. (eds) (1989) *Social Theory of Modern Societies: Anthony Giddens and his Critics.* Cambridge: Cambridge University Press.

Heller, A. (1984) *Everyday Life.* Translated by Campbell, G. London: Routledge & Kegan Paul.

Heller, A. and FehCr, F. (1991) *The Postmodern Political Condition.* Cambridge: Polity.

Heller, T., Sosna, M. and Wellbery, D. with Davidson, A., Swidler, A. and Watt, I. (eds) (1986) *Reconstructing individualism: Autonomy, individuality, and the Self in Western Thought.* Stanford, Calif.: Stanford University Press.

Heritage, J. (1984) *Garfinkel and Ethnomethodology.* Cambridge: Polity.

Heritage, J. (1990) 'Ethnomethodology'. In Giddens, A. and Turner, J. (eds).

Hill, R. J. and Stones Crittenden, K. (eds) (1968) 'Proceedings of the Purdue Symposium on Ethnomethodology'. Purdue University: Institute for the Study of Social Change, Department of Sociology.

Hindness, B. (1996) *Discourses of Power: From Hobbes to Foucault.* Oxford: Blackwell.

Hirsch, M. and Fox Keller, E. (eds) (1990) *Conflicts in Feminism.* London: Routledge.

Hole, J. and Levine, E. (1979) 'The First Feminists'. In Freeman, J. (ed.).

Hollingdale, R. J. (ed.) (1977) *A Nietzsche Reader.* Translated by Hollingdale, R. J. Harmondsworth: Penguin.

Hollis, M. and Lukes, S. (eds) (1982) *Rationality and Relativism.* Oxford: Blackwell.

Holton, R. and Turner, B. (1990) *Max Weber on Economy and Society.* London: Routledge.

Holtzman, S. and Leich, C. (eds) (1981) *Wittgenstein: To Follow a Rule.* London: Routledge & Kegan Paul.

Holub, R. C. (1991) *Jiirgen Habermas: Critic in the Public Sphere.* London: Routledge.

Homans, G. (1961) *Social Behaviour: Its Elementary Forms.* New York: Harcourt Brace Jovanovich.

Homans, G. (1973) 'Bringing Men Back In'. In Ryan, A. (ed.).

Honneth, A. (1991) *The Critique of Power: Reflective Stages in a Critical Social Theory.* Translated by Baynes, K. Cambridge, Mass.: MIT Press.

Honneth, A. (1993) 'Critical Theory in Germany Today: An Interview'. *Radical Philosophy* 65 : 33-41.

Honneth, A. and Joas, H. (1988) *Social Action and Human Nature.* Cambridge: Cambridge University Press.

Honneth, A. and Joas, H. (eds) (1991) *Communicative Action: Essays on Jiirgen Habermas's 'The Theory of Communicative Action'.* Translated by Gains, J. and Jones, D. L. Cambridge, Mass.: MIT Press.

Honneth, A., McCarthy, M., Offe, C. and Wellmer, A. (eds) (1992) *Cultural-Political Interventions in the Unfinished Project of Enlightenment.* Cambridge, Mass.: MIT Press.

Horkheimer, M. (1993) *Between Philosophy and Social Science: Selected Early Writings.* Translated by Hunter, G. F., Kramer, M. S. and Torpey, J. Cambridge, Mass.: MIT Press.

Houtkoop-Steenstra, H. (1993) 'Opening Sequences in Dutch Telephone Conversations'. In Boden, D. and Zimmerman, D. H. (eds).

Hoy, D. C. (1986a) 'Power, Repression, Progress: Foucault, Lukes, and the Frankfurt School'. In Hoy, D. C. (ed.).

Hoy, D. C. (ed.) (1986b) *Foucault: A Critical Reader.* Oxford: Blackwell.

Hoy, D. C. and McCarthy, T. (1994) *Critical Theory.* Oxford: Blackwell.

Hughes, E. (1971) *The Sociological Eye: Selected Papers.* Chicago: Aldine-Atherton.

Hughes, S. H. (1979) *Consciousness and Society: The Reorientation of European Social Thought 2* 890-1 930. Brighton: Harvester.

Humm, M. (ed.) (1992) *Feminisms: A Reader.* London: Harvester Wheatsheaf.

Jaggar, A. (1983) *Feminist Politics and Human Nature.* London: Harvester.

James, W. (1890) *Principles of Psychology.* 2 volumes. New York: Holt.

Jameson, F. (1992) *Postmodernism, or, The Cultural Logic of Late Capitalism.* London: Verso.

Jay, M. (1973) *The Dialectical Imagination: A History of the Frankfurt School and the Institute of Social Research* 1923-2 950. London: Heinemann.

Joas, H. (1985) G. *H. Mead: A Contemporary Re-Examination of his Thought.* Translated by Meyer, R. Cambridge: Polity.

Joas, H. (1993) *Pragmatism and Social Theory.* Chicago: University of Chicago Press.

Johnson, L. C. (1990) 'Socialist Feminisms'. In Gunew, S. (ed.).

Johnson, P. (1993) 'Feminism and the Enlightenment'. *Radical Philosophy* 63: 3-12.

Johnson, T., Dandeker, C. and Ashworth, C. (1984) *The Structure of Social Theory: Dilemmas and Strategies.* London: Macmillan.

Jones, C. (1992) 'Reason without Emotion'. *Radical Philosophy* 61: 32-33.

Jones, C. and Porter, R. (eds) (1994) *Reassessing Foucault: Power, Medicine and the Body.* London: Routledge.

Kamuf, P. (ed.) (1991) *A Derrida Reader: Between the Blinds.* London: Harvester Wheatsheaf.

Kaplan, E. A. and Sprinker, M. (eds) (1993) *The Althusserian Legacy.* London: Verso.

Keat, R. and Urry, J. (1975) *Social Theory as Science.* London: Routledge & Kegan Paul.

Kellner, D. (1993) 'Popular Culture and the Construction of Postmodern Identities'. In Lash, S. and Friedman, J. (eds).

Kellner, D. (ed.) (1994) *Baudrillard: A Critical Reader.* Oxford: Blackwell.

Kelly, M. (ed.) (1994) *Critique and Power: Recasting the FoucaultlHaberrnas Debate.* Cambridge, Mass.: MIT Press.

Kinnaird, J. K. (1983) 'Mary Astell: Inspired by Ideas'. In Spender, D. (ed.).

Knorr-Cetina, K. (1981) 'The Micro-Sociological Challenge of Macro-Sociology: Towards a Reconstruction of Social Theory and Methodology'. In Knorr-Cetina, K. and Cicourel, A. (eds).

Knorr-Cetina, K. and Cicourel, *A.* (eds) (1981) *Advances in Social Theory and Methodology: Towards an Integration of Micro and Macro Theories.* London: Routledge & Kegan Paul.

Kolakowski, L. (1978a) *Main Currents of Marxism, Volume 2* . Oxford: Oxford University Press.

Kolakowski, L. (1978b) *Main Currents of Marxism, Volume 2.* Oxford: Oxford University Press.

Kortian, G. (1 980) *Metacritique: The Philosophical Argument of Jiirgen Habermas.* Translated by Raffan, J. Cambridge: Cambridge University Press.

Kroker, A. and Cook, D. (1988) *The Postmodern Scene: Excremental Culture and Hyper-Aesthetics.* London: Macmillan.

Kurtz, L. (1984) *Evaluating Chicago Sociology: A Guide to the Literature with an Annotated Bibliography.* Chicago: Chicago University Press.

Laclau, E. and Mouffe, C. (1990) *Hegemony and Socialist Strategy: Towards a Radical Democratic Politics.* Third impression. London: Verso.

Landry, D. and MacLean, G. (1993) *Materialist Feminisms.* Oxford: Blackwell.

Larrabee, M. J. (ed.) (1993) *An Ethic of Care: Feminist and Interdisciplinary Perspectives.* London: Routledge.

Lash, S. (1990) *The Sociology of Postmodernism.* London: Routledge.

Lash, S. (1993) 'Pierre Bourdieu: Cultural Economy and Social Change'. In Calhoun, C., LiPuma, E. and Postone, M. (eds).

Lash, S. and Friedman, J. (eds) (1993) *Modernity and Identity.* Oxford: Blackwell.

Lash, S. and Urry, J. (1987) *The End of Organized Capitalism.* Cambridge: Polity.

Lash, S. and Urry, J. (1994) *Economies of Signs and Space.* London: Sage.

Lash, S. and Whimster, S. (eds) (1987) *Max Weber, Rationality and Modernity.* London: Allen & Unwin.

Laslett, P. and Runciman, W. G. (eds) (1967) *Philosophy, Politics and Society.* Oxford: Oxford University Press.

Lawler, S. (1995) 'Mothering the Self: A Study of the Mother-Daughter Relationship'. Unpublished Ph.D. Centre for Women's Studies, Department of Sociology, University of Lancaster.

Layder, D. (1 987) 'Key Issues in Structuration Theory: Some Critical Remarks'. *Current Perspectives in Social Theory* 8: 25-46.

Leeds Revolutionary Feminist Group (1982) 'Political Lesbianism: The Case against Heterosexuality'. In Evans, M. (ed.) .

Lemert, *C.* (1979) 'De-Centred Analysis: Ethnomethodology and Structuralism'. *Theory and Society 7:* 289-306.

Lemert, C. (ed.) (1993) *Social Theory: The Multicultural and Classic Readings.* Oxford: Westview.

Ltvi-Strauss, C. (1963) *Structural Anthropology.* New York: Basic Books.

Lewis, J. D. and Smith, R. L. (1980) *American Sociology and Pragmatism: Mead, Chicago Sociology and Symbolic Interactionism.* Chicago: University of Chicago Press.

Lincourt, J. M. and Hare, P. H. (1973) 'Neglected American Philosophers in the History of Symbolic Interactionism'. Reprinted in Plummer, K. (ed.) (1 99 1 a).

Lindsay, A. D. (1967) 'Hegel the German Idealist'. In Hearnshaw, F. J. C. (ed.).

Lockwood, D. (1976) 'Social Integration and System Integration'. In Zollschan, G. K. and Hirsch, W. (eds).

Lofland, J. (1976) Doing Social Life. New York: Wiley.

Lovell, T. (ed.) (1990) British Feminist Thought: A Reader. Oxford: Blackwell.

Lowith, K. (1993) Max Weber and Karl Marx. New edn. Preface by Turner, B. S. Edited with an introduction by Bottomore, T. and Outhwaite, W. London: Routledge.

Luhmann, N. (1982) The Differentiation of Society. Translated by Holmes, S. and Larmore, C. New York: Columbia University Press.

Luhmann, N. (1986) 'The Individuality of the Individual'. In Heller, T., Sosna, M. and Wellbery, D. with Davidson, A., Swidler, A. and Watt, I. (eds).

Luhmann, N. (1993) Risk: A Sociological Theory. Translated by Barrett, R. New York: Walter de Gruyter.

Lukiics, G. (1971) History and Class Consciousness: Studies in Marxist Dialectics. Originally published in 1923. London: Merlin.

Lukes, S. (1967) 'Alienation and Anomie'. In Laslett, P. and Runciman, W. G. (eds).

Lukes, S. (1973) 'Methodological Individualism Reconsidered'. In Ryan, A. (ed.).

Lukes, S. (1981) Emile Durkheim: His Life and Work: A Historical and Critical Study. Harmondsworth: Penguin.

Lukes, S. (1985) Marxism and Morality. Oxford: Oxford University Press.

Lyman, S. (1991) Militarism, Imperialism and Racial Accommodation: An Analysis and Interpretation of the Early Writings of Robert E. Park. Fayetteville, Ark.: University of Arkansas Press.

Lyon, D. (1994) The Electronic Eye: The Rise of Surveillance Society. Cambridge: Polity.

Lyotard, J. F. (1984) The Postmodern Condition: A Report on Knowledge. Translated by Bennington, G. and Massumi, B. Foreword by Jameson, F. Manchester: Manchester University Press.

Lyotard, J. F. (1993) 'Answering the Question: What is Postmodernism?' In Docherty, T. (ed.).

McCarthy, T. (1984) The Critical Theory of Jiirgen Habermas. Cambridge: Polity.

McCarthy, T. (1991) 'Complexity and Democracy: Or the Seducements of Systems Theory'. In Honneth, A. and Joas, H. (eds).

McHugh, P. (1974) 'On the Failure of Positivism'. In Douglas, J. D. (ed.).

McHugh, P., Raffel, S., Foss, D. C. and Blum, A. F. (1974) On the Beginning of Social Inquiry. London: Routledge & Kegan Paul.

MacIntyre, A. (1973) 'The Idea of a Social Science'. In Ryan, A. (ed.)

MacIntyre, A. (1985) After Virtue: A Study in Moral Theory. 2nd edn. London: Duckworth.
MacKinnon, C. (1988) 'Desire and Power: A Feminist Perspective'. In Nelson, C. and Grossberg, L. (eds).
McLellan, D. and Sayers, S. (eds) (1990) Socialism and Morality. London: Macmillan.
McLennan, G. (1995) Pluralism. Buckingham: Open University Press.
McNay, L. (1994) Foucault: A Critical Introduction. Cambridge: Polity.
McPhail, C. and Rexroat, C. (1979) 'Mead vs. Blumer: The Divergent Methodological Perspectives of Social Behaviourism and Symbolic Interactionism'. Reprinted in Plummer, K. (ed.) (1991a).
Mannheim, K. (1952) *Essays on the Sociology of Knowledge.* Translated and edited by Kecskemeti, P. New York: Oxford University Press.
Manning, P. (1992) *Erving Goffman and Modern Sociology.* Cambridge: Polity.
Marcuse, H. (1966) *Eros and Civilization: A Philosophical Inquiry into Freud.* 2nd edn. Boston, Mass.: Beacon.
Marcuse, H. (1968a) *One Dimensional Man: The Ideology of Industrial Society.* London: Sphere.
Marcuse, H. (1 968b) *Negations: Essays in Critical Theory.* Boston, Mass.: Beacon.
Marcuse, H. (1969) *Reason and Revolution: Hegel and the Rise of Social Theory.* London: Routledge & Kegan Paul.
Marshall, B. L. (1994) *Engendering Modernity: Feminism, Social Theory and Social Change.* Cambridge: Polity.
Martin, J. R. (1994) 'Methodological Essentialism, False Difference, and Other Dangerous Traps'. *Signs* 19(3): 630-657.
Martin, L. H., Gutman, H. and Hutton, P. H. (eds) (1988) *Technologies of the Self: A Seminar with Michel Foucault.* London: Tavistock.
Marx, K. (1961) *The Poverty of Philosophy.* Originally published in 1847. Edited by Arthur, C. J. London: Lawrence & Wishart.
Marx, K. (1964) *The German Ideology.* London: Lawrence & Wishart.
Marx, K. (1980) 'The Eighteenth Brumaire of Louis Bonaparte'. In Marx, K. and Engels, F. Marx, K. (1981) *Economic and Philosophical Manuscripts of 1844.* London: Lawrence & Wishart.
Marx, K. (1983) *Capital: A Critique of Political Economy. Volume 2 : The Process of Production of Capital.* Edited by Engels, F. Translated by Moore, S. and Aveling, E. English edition first published in 1887. London: Lawrence & Wishart.

Marx, K. and Engels, F. (1953) *Selected Correspondence: 1843-2895.* London: Lawrence & Wishart.

Marx, K. and Engels, F. (1970) *The German Ideology.* London: Lawrence & Wishart.

Marx, K. and Engels, F. (1980) *Selected Works in One Volume.* London: Lawrence & Wishart.

Mayer, T. (1994) *Analytical Marxism.* London: Sage.

Maynard, D. W. and Clayman, S. E. (1991) 'The Diversity of Ethnomethodology'. *Annual Review of Sociology* 17: 385-41 8.

Mead, G. H. (1 9 10) 'What Social Objects Must Psychology Presuppose?' Reprinted in Reck, A. J. (ed.).

Mead, G. H. (1930) 'Cooley's Contribution to American Social Thought'. Reprinted in Plummer, K. (ed.) (1991a).

Meltzer, B. N. and Petras, J. W. (1973) 'The Chicago and Iowa Schools of Symbolic Interactionism'. In Shibutani, T. (ed.).

Meltzer, B. N., Petras, J. W. and Reynolds, L. T. (1975) *Symbolic Interactionism: Genesis, Varieties and Criticism.* London: Routledge & Kegan Paul.

Mennell, S. (1975) 'Ethnomethodology and the New *Methodenstreit'. Acta Sociologica* 18(4): 287-302.

Mennell, S. (1992) *Norbert Elias: An Introduction.* Oxford: Blackwell.

Merton, R. (1968) *Social Theory and Social Structure.* New York: Free Press.

MeStrovii, S. (1992) *The Coming Fin de Sidcle: An Application of Durkheim's Sociology to Modernity and Postmodernity.* London: Routledge.

MeBtroviC, S. (1993) The Barbarian Temperament: Towards a Critical Postmodern Theory. London: Routledge.

Milband, R. (1990) 'Class Analysis'. In Giddens, A. and Turner, J. (eds).

Millett, K. (1969) Sexual Politics. New York: Avon.

Millett, K. (1992) 'Beyond Politics? Children and Sexuality'. In Vance, C. S. (ed.).

Mills, C. W. (1970) The Sociological Imagination. Originally published in 1959. Harmondsworth: Penguin.

Mitchell, J. (1966) 'Women: The Longest Revolution'. New Left Review 40: 11-37.

Mitchell, J. (1 974) Woman's Estate. Harmondsworth: Penguin.

Mitchell, J. (1986) 'Reflections on Twenty Years of Feminism'. In Mitchell, J. and Oakley, A. (eds).

Mitchell, J. and Oakley, A. (eds) (1986) What is Feminism? Oxford: Blackwell.

Moi, T. (1 99 1) 'Appropriating Bourdieu: Feminist Theory and Pierre Bourdieu's Sociology of Culture'. New Literary History 22: 1017-1049.

Mommsen, W. J. (1992) The Political and Social Theory of Max Weber: Collected Essays. Cambridge: Polity.

Monk, R. (1990) Ludwig Wittgenstein: The Duty of Genius. London: Vintage.

Mouzelis, N. (1991) Back to Sociological Theory: The Construction of Social Orders. London: Macrnillan.

Mouzelis, N. (1995) Sociological Theory: What Went Wrong? Diagnoses and Remedies. London: Routledge.

Mullan, B. (1987) Sociologists on Sociology: Interviews with Contemporary Sociologists. London: Croom Helm.

Murgatroyd, L. (1989) 'Only Half the Story: Some Blinkering Effects of 'Malestream" Sociology'. In Held, D. and Thompson, J. B. (eds).

Nelson, C. and Grossberg, L. (eds) (1988) Marxism and the Interpretation of Culture. London: Macmillan.

Nisbet, R. (1970) The Sociological Tradition. London: Heinemann.

Norris, C. (1987) Derrida. London: Fontana.

Norris, C. (1996) Reclaiming Truth: Contribution to a Critique of Cultural Relativism. London: Lawrence & Wishart.

Nuyen, A. T. (1990) 'Truth, Method, and Objectivity: Husserl and Gadamer on Scientific Method'. Philosophy of the Social Sciences 20(4): 437452.

Oakley, A. (1972) Sex, Gender and Society. London: Maurice Temple Smith.

Oakley, A. (1974) The Sociology of Housework. Oxford: Martin Robertson.

Oakley, A. and Mitchell, J. (eds) (1976) The Rights and Wrongs of Women. Harmondsworth: Penguin.

O'Connor, J. (1987) The Meaning of Crisis: A Theoretical Introduction. Oxford: Blackwell.

Offe, C. (1985) Disorganized Capitalism. Edited by Keane, J. Cambridge: Polity.

Okin, S, M. (1980) Women in Western Political Thought. London: Virago.

Ollman, B. (1979) Social and Sexual Revolution: Essays on Marx and Reich. London: Pluto.

O'Neill, J. (1995) The Poverty of Postmodernism. London: Routledge.

Ortner, S. B. (1982) 'Is Female to Male as Nature is to Culture?' In Evans, M. (ed.).

Osborn, R. (1937) Freud and Marx: A Dialectical Study. Introduction by Strachey, J. London: Gollancz.

Outhwaite, W. (1975) Understanding Social Life: The Method Called 'Verstehen'. London: Allen & Unwin.

Outhwaite, W. (1994) *Habermas*. Cambridge: Polity.

Paci, E. (1 972) *The Function of the Sciences and the Meanings of Man*. Translated by Piccone, P. and Hansen, J. Evanston, Ill.: Northwestern University Press.

Park, R. E. (1936) 'Human Ecology'. *American Journal of Sociology* 42: 1-15.

Park, R. E. (1972) *The Crowd and the Public and Other Essays*. Edited by Elsner, H. Translated by Elsner, C. Chicago: University of Chicago Press.

Parsons, T. (1937) *The Structure of Social Action*. New York: McGraw-Hill.

Parsons, T. (1951) *The Social System*. New York: Free Press.

Parsons, T. (1 965) 'An Outline of the Social System'. In Parsons, T., Shils, E., Naegele, K. D. and Pitts, J. R. (eds).

Parsons, T., Shils, E., Naegele, K. D. and Pitts, J. R. (eds) (1965) *Theories of Society: Foundations of Modern Sociological Theory*. New York: Free Press.

Pearce, F. (1989) *The Radical Durkheim*. London: Unwin Hyman.

Petras, J. W. (1968) 'Psychological Antecedents of Sociological Theory in America: Williams James and James Mark Baldwin'. Reprinted in Plummer, K. (ed.) (1991a).

Plamenatz, J. (1963) *Man and Society: Volume 1: A Critical Examination of Some Important Social and Political Theories from Machiavelli to Marx*. London: Longman.

Plummer, K. (ed.) (1 991 a) *Symbolic Interactionism, Volume 1 : Foundations and History*. Aldershot: Edward Elgar.

Plummer, K. (ed.) (1 991 b) *Symbolic Interactionism, Volume 2: Contemporary Issues*. Aldershot: Edward Elgar.

Pollner, M. (1991) 'Left of Ethnomethodology: The Rise and Decline of Radical Reflexivity'. *American Sociological Review* 56(3): 370-380.

Poulantzas, N. (1 978) *State, Power, Socialism*. Translated by Camiller, P. London: New Left Books.

Ramazanoglu, C. (1990) *Feminism and the Contradictions of Oppression*. London: Routledge.

Ramazanoglu, C. (ed.) (1993) *Up against Foucault: Explorations of some Tensions between Foucault and Feminism*. London: Routledge.

Rattansi, A. (1 994) ' "Western" Racisms, Ethnicities and Identities in a " Postmodern" Frame'. In Rattansi, A. and Westwood, S. (eds).

Rattansi, A. and Westwood, S. (1994a) 'Modern Racisms, Racialized Identities'. In Rattansi, A. and Westwood, S. (eds).

Rattansi, A. and Westwood, S. (eds) (1994b) *Racism, Modernity and Identity on the Western Front*. Cambridge: Polity.

Rawls, A. W. (1987) 'The Interaction Order *Sui Generis:* Goffman's Contribution to Social Theory. *Sociological Theory* 5 (2) : 13 6-149.
Ray, L. and Reed, M. (eds) (1 994) *Organizing Modernity: New Weberian Perspectives on Work, Organization and Society.* London: Routledge.
Reck, A. J. (ed.) (1964) *Selected Writings: George Herbert Mead.* Chicago: University of Chicago Press.
Rex, J. (1 961) *Key Problems ofSociological Theory.* London: Routledge & Kegan Paul.
Rich, A. (1 980) 'Compulsory Heterosexuality and Lesbian Existence'. *Signs: Journalof Women in Culture and Society* 5 *(4:)* 631 -660.
Ricoeur, P. (1982) *Hermeneutics and the Human Sciences.* Edited and translated by Thompson, J. B. Cambridge: Cambridge University Press.
Risseeuw, C. (1991) 'Bourdieu, Power and Resistance: Gender Transformation in Sri Lanka'. In Davis, K., Leijenaar, M. and Oldersma, J. (eds).
Robbins, D. (1991) *The Work of Pierre Bourdieu: Recognizing Society.* Milton Keynes: Open University Press.
Robertson, R. and Turner, B. S. (eds) (1991) *Talcott Parsons: Theorist of Modernity.* London: Sage.
Rocher, G. (1974) *Talcott Parsons and American Sociology.* Translated by Mennell, B. and Mennell, S. London: Thomas Nelson.
Rock, P. (1979) *The Making of Symbolic Interactionism.* London: Macmillan.
Rogers, K. M. (ed.) (1979) *Before Their Time: Six Women Writers of the Eighteenth Century.* New York: Frederick Unger.
Rogers, M. F. (1977) 'Goffman on Power'. *American Sociologist* 12(April): 88-95.
Rogers, M. F. (1983) *Sociology, Ethnomethodology, and Experience: A Phenomenological Critique.* Cambridge: Cambridge University Press.
Rogers, M. F. (1984) 'Everyday Life as Text'. In Collins, R. (ed.).
Rorty, R. (1982) *Consequences of Pragmatism: Essays 1972-1* 980. Minneapolis, Minn.: University of Minneapolis Press.
Rorty, R. (1 989) *Contingency, Irony and Solidarity.* Cambridge: Cambridge University Press.
Rorty, R. (1993) 'Cosmopolitanism without Emancipation: A Response to Lyotard'. In Lash, S. and Friedman, J. (eds).
Rose, A. (ed.) (1972) *Human Behaviour and Social Processes: An Interactionist Approach.* London: Routledge & Kegan Paul.
Rose, J. (1990) 'Femininity and its Discontents'. In Lovell, T. (ed.).
Ross, D. (1991) *The Origins of American Social Science.* Cambridge: Cambridge University Press.

Rowbotham, S. (1973) *Woman's Consciousness, Man's World.* Harmondsworth: Penguin.

Rowland, R. and Klein, R. D. (1990) 'Radical Feminism: Critique and Construct'. In Gunew, S. (ed.).

Rubinstein, D. (1981) *Marx and Wittgenstein: Social Praxis and Social Explanation.* London: Routledge & Kegan Paul.

Runciman, W. G. (1973) 'What is Structuralism?' In Ryan, A. (ed.).

Russell, B. (1955) *History of Western Philosophy and its Connection with Political and Social Circumstances from the Earliest Times to the Present Day.* Fifth impression. London: Allen & Unwin.

Ryan, A. (ed.) (1973) *The Philosophy of Social Explanation.* Oxford: Oxford University Press.

Ryan, A. (1986) *Property and Political Theory.* Oxford: Blackwell.

Sacks, H. (1963) 'Sociological Description'. *Berkeley Journal of Sociology* 8: 1-16.

Sacks, H. (1972) 'An Initial Investigation of the Usability of Conversational Data for Doing Sociology'. In Sudnow, D. (ed.).

Sartre, J.-P. (1960) *Critique of Dialectic Reason.* London: New Left Books.

Sarup, M. (1993) *An Introductory Guide to Post-Structuralism and Postmodernism.* 2nd edn. London: Harvester Wheatsheaf.

Sayer, A. (1992) *Method in Social Science: A Realist Approach.* 2nd edn. London: Routledge.

Sayer, D. (1991) *Capitalism and Modernity: An Excursus on Marx and Weber.* London: Routledge.

Sayers, S. (1989) 'Knowledge as Social Phenomenon'. *Radical Philosophy* 52: 34-37.

Schmidt, E. P. (ed.) (1937) *Man and Society.* New York: Prentice Hall.

Schroeder, R. (1 992) *Max Weber and the Sociology of Culture.* London: Sage.

Schutz, A. (1972) *The Phenomenology of the Social World.* Originally published in 1932. Translated by Walsh, G. and Lehnert, F. Introduction by Walsh, G. London: Heinemann.

Schutz, A. (1979) 'Concept and Theory Formation in the Social Sciences'. In Bynner, J. and Stribley, K. (eds).

Schwartz, J. (1990) 'Antihumanism in the Humanities'. *Public Interest* 99: 29-44.

Sciulli, D. (1985) 'The Practical Groundwork of Critical Theory: Bringing Parsons to Habermas (and vice versa)'. In Alexander, J. (ed.).

Segal, L. (1 987) *Is the Future Female? Troubled Thoughts on Contemporary Feminism.* London: Virago.

Seidman, S. (1994a) 'The End of Sociological Theory'. In Seidman, S. (ed.) (1994~).

Seidman, S. (1994b) *Contested Knowledge: Social Theory in the Postmodern Era.* Oxford: Blackwell.

Seidman, S. (ed.) (1 994c) *The Postmodern Turn: New Perspectives on Social Theory.* Cambridge: Cambridge University Press.

Seidman, S. and Wagner, D. (eds) (1992) *Postmodernism and Social Theory.* Oxford: Blackwell.

Sharrock, W. and Anderson, B. (1986) *The Ethnomethodologists.* London: Tavistock.

Sharrock, W. and Anderson, B. (1991) 'Epistemology: Professional Scepticism'. In Button, G. (ed.).

Shibutani, T. (ed.) (1 973) *Human Natureand Collective Behaviour: Papers in Honor of Herbert Blumer.* New Brunswick, NJ: Transaction.

Shilling, C. (1993) *The Body and Social Theory.* London: Sage.

Shott, S. (1987) 'Emotion and Social Life: A Symbolic Interactionist Analysis'. Reprinted in Plummer, K. (ed.) (1 99 1 b).

Shotter, J. (1 993) *Cultural Politics of Everyday Life: Social Constructionism, Rhetoric and Knowing of the Third Kind.* Buckingham: Open University Press.

Simons, H. W. and Billig, M. (eds) (1994) *After Postmodernism: Reconstructing Ideology Critique.* London: Sage.

Simons, J. (1995) *Foucault and the Political.* London: Routledge.

Singer, P. (1983) *Hegel.* Oxford: Oxford University Press.

Skinner, Q. (ed.) (1991) *The Return of Grand Theory in the Human Sciences.* Cambridge: Cambridge University Press.

Slater, D. (1994) 'Exploring Other Zones of the Postmodern: Problems of Ethnocentrism and Difference across the North-South Divide'. In Rattansi, A. and Westwood, S. (eds).

Smart, B. (1990) 'Modernity, Postmodernity and the Present'. In Turner, B. S. (ed.).

Smith, C. W. (1983) 'A Case Study of Structuration: The Pure-Bred Beef Business'. *Journal for the Theory of Social Behaviour* 13 : 4-1 8.

Smith, D. E. (1 974) 'Theorizing as Ideology'. In Turner, R. (ed.).

Smith, D. E. (1 987) *The Everyday World as Problematic: A Feminist Sociology.* Milton Keynes: Open University Press.

Smith, D. E. (1993) *Texts, Facts and Femininity: Exploring the Relations of Ruling.* London: Routledge.

Smith, N. (1993) 'Putting Practice into Theory: A Review Essay'. *Radical Philosophy* 63: 4244.

Soper, K. (1993) 'Productive Contradictions'. In Ramazanoglu, C. (ed.).

Spencer, H. (1969) *Herbert Spencer.* Edited with an introduction by Low-Beer, A.London: Macmillan.

Spender, D. (1980) *Man-Made Language.* London: Routledge & Kegan Paul.

Spender, D. (ed.) (1983) *Feminist Theorists: Three Centuries of Women's Intellectual Traditions.* London: Women's Press.

Spender, D. (1983) *Women of Ideas (and what men have done to them).* London: Ark.

Spybey, T. (1995) *Globalization and World Society.* Cambridge: Polity.

Stinchcombe, A. L. (1990) 'Milieu and Structure Updated: A Critique of the Theory of Structuration'. In Clark, J., Modgil, C. and Modgil, S. (eds).

Strasser, S. (1985) *Understanding and Explanation: Basic Ideas Concerning the Humanity of the Human Sciences.* Pittsburgh, Pa: Duquesne University Press.

Strauss, A. (1978) *Negotiations, Varieties, Contexts, Processes and Social Order.* San Francisco: Jossey Bass.

Strauss, A. (1982) 'Interorganizational Negotiation'. Reprinted in Plummer, K. (ed.) (1991b).

Stryker, S. (1980) *Symbolic Interactionism: A Social Structural Version.* Menlo Park, Calif.: BenjaminICummings.

Sudnow, D. (ed.) (1972) *Studies in Social Interaction.* New York: Free Press.

Sydie, R. (1987) *Natural Women, Cultured Men: A Feminist Perspective on Sociological Theory.* Milton Keynes: Open University Press.

Tannen, D. (ed.) (1993) *Gender and Conversational Interaction.* New York: Oxford University Press.

Taylor, C. (1981) 'Understanding in Human Science'. *Review of Metaphysics* 34: 25-38.

Taylor, C. and Montefiore, A. (1980) 'From an Analytical Perspective'. An introductory essay in Kortian, G.

Thomas, W. I. and Thomas, D. S. (1928) *The Child in America: Behavior Problems and Programs.* New York: Knopf.

Thomas, W. I. and Znaniecki, F. (1918-19) *The Polish Peasant in Europe and America.* 5 volumes. Boston, Mass.: Badger.

Thompson, J. B. (1981) *Critical Hermeneutics:* A *Study in the Thought of Paul Ricoeur and Jiirgen Habermas.* Cambridge: Cambridge University Press.

Thompson, J. B. (1 984) *Studies in the Theory of Ideology.* Cambridge: Polity.

Thorne, B., Kramerae, C. and Henley, N. (eds) (1983) *Language, Gender and Society.* Rowley, Mass.: Newbury House.

Tiryakian, E. A. (1979) 'Emile Durkheim'. In Bottomore, T. and Nisbet, R. (eds).

Tong, R. (1989) *Feminist Thought: A Comprehensive Introduction.* London: Unwin Hyman.

Toynbee, A. (1954) *A Study of History: Volume 9.* London: Oxford University Press.

Turner, B. S. (1 98 1) *For Weber: Essays on the Sociology of Fate.* London: Routledge & Kegan Paul.

Turner, B. S. (ed.) (1990) *Theories of Modernity and Postmodernity.* London: Sage.

Turner, B. S. and Robertson, R. (1991) 'How to Read Parsons'. In Robertson, R. and Turner, B. S. (eds).

Turner, J. H. (1974) 'Parsons as a Symbolic Interactionist: A Comparison of Action and Interaction Theory'. Reprinted in Plummer, K. (ed.) (1 99 1 a).

Turner, R. (1974a) 'Words, Utterances and Activities'. In Turner, R. (ed.).

Turner, R. (ed.) (1974b) *Ethnomethodology.* Harmondsworth: Penguin.

Vance, C. S. (ed.) (1992) *Pleasure and Danger: Exploring Female Sexuality.* London: Pandora.

Visker, R. (1995) *Michel Foucault: Genealogy as Critique.* Translated by Turner, C. London: Verso.

Wacquant, L. (1992) 'The Structure and Logic of Bourdieu's Sociology'. In Bourdieu, P. and Wacquant, L.

Wacquant, L. (1993) 'Bourdieu in America: Notes on the Transatlantic Importation of Social Theory'. In Calhoun, C., LiPuma, E. and Postone, M. (eds).

Walby, S. (1986) *Patriarchy at Work: Patriarchal and Capitalist Relations in Employment.* Cambridge: Polity.

Walby, S. (1990) *Theorizing Patriarchy.* Cambridge: Polity.

Wallerstein, I. (1990) 'World-Systems Analysis'. In Giddens, A. and Turner, J. (eds).

Warren, M. (1992) 'Max Weber's Nietzschean Conception of Power'. *History of the Human Sciences 5* (3) : 19-3 7.

Weber, M. (1949) *The Methodology of the Social Sciences.* Edited by Shils, E. and Finch, H. Glencoe, Ill.: Free Press.

Weber, M. (1985) *The Protestant Ethic and the Spirit of Capitalism.* Originally published in 1930. London: Unwin.

Webster, F. (1 995) *Theories of the Information Society.* London: Routledge.

Weedon, C. (1987) *Feminist Practice and Poststructuralist Theory*. Oxford: Blackwell.

Weinbaum, B. (1978) *The Curious Courtship of Women's Liberation and Socialism.* Boston, Mass.: South End Press.

Weinstein, D. and Weinstein, M. (1991) 'Georg Simmel: Sociological Fliineur Bricoleur'. In Featherstone, M. (ed.).

West, C. and Zimmerman, D. H. (1983) 'Small Insults: A Study of Interruptions in Cross-Sex Conversations with Unacquainted Persons'. In Thorne, B., Kramerae, C. and Henley, N. (eds).

White, S. (1990) *The Recent Work of Jiirgen Habermas: Reason, Justice and Modernity*. Cambridge: Cambridge University Press.

Wieder, D. L. (1 974) 'On Meaning by Rule'. In Douglas, J. D. (ed.).

Wiggershaus, R. (1995) *The Frankfurt School: Its History, Theories and Political Significance.* Translated by Robertson, M. Cambridge: Polity.

Williams, M. and May, T. (1996) *Introduction to the Philosophy of Social Research*. London: UCL Press.

Williams, Raymond (1 961) *Culture and Society: 1 780-1 950.* Harmondsworth: Penguin.

Williams, Robin (1980) 'Goffman's Sociology of Talk'. In Ditton, J. (ed.).

Williams, Robin (1988) 'Understanding Goffman's Methods'. In Drew, P. and Wootton, A. (eds).

Wilson, B. (ed.) (1970) *Rationality*. Oxford: Blackwell.

Wilson, E. (1990) 'Psychoanalysis: Psychic Law and Order?' In Lovell, T. (ed.).

Winch, P. (1990) The Idea of a Social Science and its Relation to Philosophy. 2nd edn. Originally published in 1958. London: Routledge.

Wolff, K. H. (1979) 'Phenomenology and Sociology'. In Bottomore, T. and Nisbet, R. (eds).

Wolffensperger, J. (1991) 'Engendered Structure: Giddens and the Conceptualisation of Gender'. In Davis, K., Leijenaar, M. and Oldersma, J. (eds).

Wollstonecraft, M. (1989) A Vindication *of* the Rights of Women. Originally published in 1792. Buffalo, NY: Prometheus.

Wright, E. 0. (1989) 'Models of Historical Trajectory: An Assessment of Giddens's Critique of Marxism'. In Held, D. and Thompson, J. B. (eds).

Zimmerman, D. H. and Wieder, D. L. (1974) 'Ethnomethodology and the Problem of Order: Comments on Denzin'. In Douglas, J. D. (ed.).

Zollschan, G. K. and Hirsch, W. (eds) (1976) Social Change: Explorations, Diagnoses and Conjectures. New York: Wiley.

INDEX

#

A

B

C

F

G

H

I

O

P

Q

R